SCIENCE FUNDAMENTALS:

A BACKGROUND FOR HOUSEHOLD EQUIPMENT

SCIENCE FUNDAMENTALS:
A BACKGROUND FOR HOUSEHOLD EQUIPMENT

Louise J. Peet

The Iowa State University Press, Ames, Iowa

LOUISE J. PEET is Professor of Family Environment, Iowa State University. For twenty-two years she was professor and head of the department of household equipment; before this appointment she taught at William Woods College, Fulton, Missouri, and Istanbul Women's College in Turkey. She has been also visiting professor at the University of Illinois, the University of Wyoming, Oregon State College, Brigham Young University, and Utah State University. She holds the BA and the MA degrees from Wellesley College and a doctorate from Iowa State University.

Besides this work, her publications include: *Household Equipment*, 6th ed., *Young Homemakers' Equipment Guide*, and numerous articles on equipment in the journals of her field. She has received the Faculty Award of Iowa State University and was given the Alumni Merit Award of the Iowa State Club of Chicago in 1970.

Composed and printed by
The Iowa State University Press

First edition, 1972

Library of Congress Cataloging in Publication Data

Peet, Louise (Jenison) 1885–
Science fundamentals.

Bibliography: p.
1. Science. 2. Home economics–Equipment and supplies. I. Title.
TX149.P43 530'.02'4649 70–171171
ISBN 0–8138–0835–9

CONTENTS

PREFACE

INFORMATION in household equipment is based largely on a fundamental knowledge of chemistry and physics. Structure of matter, materials used in the construction and finishing of appliances and utensils, and the various laundry aids are usually considered chemical in nature while the other subjects covered in this book—electricity, electronics, light, gas, heat, mechanics, and the items considered under pollutants—have a background of the subject-matter covered in the discipline of physics.

At the present time material contained in conventional textbooks of chemistry and physics tends to be highly mathematical in nature, and unless the teacher of equipment is well grounded in mathematics, much of the information will be difficult to understand. Therefore, it has been my intention to provide a text with necessary background material conveyed in simple terms.

The information in this book has been selected from the author's extensive readings in recent texts on chemistry and physics. However, the chapter on gas and the chapter on pollutants, household and environmental, contain certain information not generally found in the usual textbook. Lists of magazines, manufacturers, and associations from which up-to-date information can be obtained are included in the bibliography of Research Sources.

Two physics texts of particular value are N. C. Harris and E. M. Hemmerling, *Introductory Applied Physics,* 2d ed. (1963) and L. T. Earls, *Brief Course in Physics* (1968).

The manuscript has been read for accuracy of technical information by Professor Harvey Diehl of the Chemistry Department, and by professors Percy Carr and Herbert Plagge of the Physics Department of Iowa State University. The chapter on gas was read by W. F. Lariviere of the Iowa Electric Light and Power Company, Ames, Iowa. To these gentlemen the author expresses her very great appreciation.

Some of the illustrations in this book are the standard ones found in numerous textbooks; others have been supplied by the commercial companies in the field.

SCIENCE FUNDAMENTALS:
A BACKGROUND FOR HOUSEHOLD EQUIPMENT

1 STRUCTURE OF MATTER

APPLIANCES and utensils are made from stainless steel, aluminum, iron, glass, tinned steel, and the various plastics, all of which are chemical substances. The materials used in laundry procedures: water, detergents, bleaches, starches, fabric softeners, and others, are also chemical in nature, made up of elements or compounds whose interactions may be expressed in equations. These chemical substances comprise matter.

PARTICLES. Matter is made of many exceedingly small particles called atoms and molecules, and it exists in three states, solid, liquid, and gas. These three states differ basically in the position of the particles relative to one another. In solids, the particles vibrate about fixed points and are held in their relative positions by cohesive forces; in liquids, they are not so tightly held and may change positions by sliding across one another; in gases the particles are much farther apart and tend to occupy all the space of the container which encloses them. Particles usually vibrate with extremely high velocities, the velocity measured in kilometers per second. The number of kilometers depends on the temperature; more kilometers at a high temperature, less at a lower temperature.

ELEMENTS. According to Dalton's Theory, first proposed in the early nineteenth century, a particle is made of even smaller basic particles. Ninety-two different kinds of basic particles, known as elements, are commonly found in nature, and a few others have been synthesized artificially. Each element is unique in mass and properties. Basic particles combine to form countless compounds, and new ones are discovered from time to time. The molecule is the smallest particle of a compound; the atom, the smallest particle of an element.

Symbols. For ease in expressing the composition of compounds, the elements are written symbolically. The symbols for some elements have been determined from the name of the element and the first or first two letters of the name used, for example: H for hydrogen, O for oxygen, Al for aluminum, and Ca for calcium. Other symbols have been derived from the Latin or Greek name of the element: Cu for copper *(cuprum)*, Au for gold *(aureum)*, Ag for silver *(argentum)*, and Na for sodium *(natrium)* were all derived from Latin. Among the symbols derived from the Greek are Neon (Ne), Krypton (Kr), Xenon (Xe), Helium (He), and Thallium (Tl).

Atomic Weight. The atomic weights of the elements were first determined by using the mass of hydrogen, the lightest element, as a basis, and the masses of the other elements were expressed in relation to the hydrogen mass of 1.0000. Later, the mass of oxygen, with a value of 16.0000, was used. When it was discovered that elements have isotopes, and that many elements exist in several forms differing only slightly in mass, one of the isotopes of carbon, carbon 12 with an atomic value of 12.0000, was adopted internationally as the standard to which the values of the other elements could be compared. Using C 12.0000 as a base, oxygen has an atomic weight of 15.9994 instead of 16.0000, and hydrogen, 1.00797 instead of 1.0000. Such minute differences are important only to the scientist making very exact measurements.

During chemical reactions, atoms form different group arrangements, making new molecules of different chemical composition. The individual atoms do not gain or lose mass; therefore, when forming compounds, the atoms of the various elements always combine in definite proportions. When the molecules retain their original identities, a mixture is formed. Air, for example, is a mixture of nitrogen (N_2), oxygen (O_2), carbon dioxide (CO_2), and water vapor (H_2O).

CHARACTERISTICS OF ATOMS. Atoms are spherical in shape. The radius of an atom, the straight-line distance from its center to the surface of the sphere, is usually considered to be from 0.1 to 0.2 millimicrons. The space within the sphere of the atom is largely empty, like our own solar system. And like the solar system the atom has at its center a nucleus around which electrons revolve in orbits at varying distances from the nucleus. Some of these orbits are circular; others are elliptical in form.

PROTONS AND NEUTRONS. The nucleus is composed of positively charged particles called protons and uncharged particles, neutrons. The number of neutrons approximates the number of protons in the nucleus, and the two are about equal in mass. The number of protons is the same for all the isotopes of a given element, but the number of neutrons varies from isotope to isotope, causing a difference in the atomic weight. Since the number of protons remains the same, the chemical characteristics of a series of isotopes of an element will also be the same.

When the revolving electrons carry negative charges equal in number to the positive charges the atom is known as a neutral atom, and the electrical attractions between the electrons and the protons keep the electrons in their orbits. The combining properties of the atoms result from the interaction of the outer electrons.

In certain metals, aluminum and copper for example, the electrons in the outer orbits are not held to the nucleus too tenaciously and, by the force of voltage, may be made to move from one atom to another producing an electric current which flows through the metal.

THE PERIODIC TABLE. The number of electrons in an atom is its atomic number and varies from 1 for hydrogen to 92 for uranium. Thus, the atomic radius of uranium is far greater than that of hydrogen. The subject is complex, however, because the atomic radius goes up and down in each horizontal row of the periodic system.

When several of the early chemical scientists arranged the elements according to their atomic weights, they found that certain elements had similar characteristics and could be placed under one another in vertical columns in a kind of family. For example, lithium, sodium, potassium, and rhubidium form one such group; magnesium, calcium, strontium, and barium another group; and fluorine, chlorine, bromine, and iodine a third. The series of vertical columns, eight in all, and each divided into A and B groups based on differences in properties, form the Periodic Table. With the discovery of electrons, the periodic system was rearranged according to the atomic numbers of the atoms. This change did not alter the positions of the various groupings.

Mendeleev, a Russian chemist, was the author of the early periodic table most widely accepted. He left empty spaces in his arrangement and predicted that other elements would be discovered at some future time to fit into these spaces. This prediction inspired investigators to search for the missing elements and led to the discovery of a number of new ones. An element is usually more like the other elements in its own family than it is like elements in other families. It also tends to form similar compounds.

While the elements in the vertical columns are similar, differences exist between the elements in the horizontal rows and this accounts for the variations in atomic radius.

PROJECTS AND EXPERIMENTS

1. To determine the three states of water:
   a. Examine an ice cube and allow it to melt at room temperature. Note the time required, and compare the volume of the liquid with the volume of the cube.
   b. Heat the liquid in a stoppered flask and note the time required for the steam to form. Allow to cool. How much condensation has taken place?
   c. Freeze an equal volume of liquid in a small flask or bottle, again noting time.

2. List the symbols of 20 elements with which you are familiar, and note whether the symbol has been derived from the English, Latin, or Greek name of the element.

A B	A B	A B	A B	A B	A B	A B	A B		
1 H .0080							2 He		
3 Li	4 Be	5 B	6 C	7 N	8 O	9 F	10 Ne		
11 Na	12 Mg	13 Al	14 Si	15 P	16 S	17 Cl	18 Ar		
19 K	20 Cz	21 Sc	22 Ti	23 V	24 Cr	25 Mn	26 Fe	27 Co	28 Ni
29 Cu	30 Zn	31 Ga	32 Ge	33 As	34 Se	35 Br	36 Kr		
37 Rb	38 Sr	39 Y	40 Zr	41 Nb	42 Mo	43 Tc	44 Ru		
47 Ag	48 Cd	49 In	50 Sn	51 Sb	52 Te	53 I	54 Xe		
55 Cs	56 Ba	57 La	72 Hf	73 Ta	74 W	75 Re	76 Ds		
79 Au	80 Hg	81 Tl	82 Pb	83 Bi 208.98	84 Po	85 At	86 Rn		

FIG. 1.1. Short Form of Periodic Table. All the original elements are included and some of the later-discovered members. Elements are arranged according to their periodic physical and chemical similarities in the same vertical A or B groups. Above each element is its atomic number. The atomic weights are given for hydrogen and bismuth. The atomic weights of the other elements lie between these two numbers.

2 MATERIALS USED IN HOUSEHOLD EQUIPMENT

FOR THE INTELLIGENT BUYING of equipment used in the kitchen and laundry, the homemaker must have some basic knowledge of the properties of the materials from which the appliances and utensils are made as well as the finishes that make the equipment resistant to corrosion and easy to clean. Such knowledge makes it possible to select the most desirable material for a given task, use it successfully, and care for it so that it will last a long time, eliminating the necessity of time wasted and money spent in its replacement.

Materials used in household equipment may be divided into three groups: (1) those used in appliances and utensils and their finishes; (2) those used for insulation; and (3) those used for cabinet construction and wall, floor, and counter coverings.

MATERIALS USED IN APPLIANCES AND UTENSILS

METALS. Many different metals are used to construct household appliances and utensils from ranges, refrigerators, dishwashers, clothes washers and dryers to the pots and pans used in surface and oven cookery. Small electric equipment such as toasters, grills, unit ovens, coffee makers, frypans, and corn poppers as well as numerous tools needed in food preparation are also made of metal. While various plastics are popular for certain uses, they supplement rather than replace the metals.

In the early construction of coal and wood ranges, cast iron was used almost entirely, although not always satisfactorily. Cast iron is heavy and brittle, and the ranges were constructed by bolting the cast pieces together. When the ranges had to be moved, the bolts tended to loosen and breakage often resulted. Sheet iron was used for the frames of oil stoves, but it too was unsatisfactory as it tended to warp when subjected to high heat. Now, low-carbon steel is employed almost exclusively for the frames of ranges as well as refrigerators, washers, and dryers. The steel has many desirable characteristics: it is comparatively light in weight and, most important of all, it can be welded so that the parts become a complete whole with no cracks or crevices through which air can circulate. A welded frame is especially desirable for the range oven and the refrigerator, both of which have double-walled construction with insulation between the walls, and moisture present in their interiors. Because moisture is a good conductor of heat, it must be kept out of the insulation. We want the heat to remain in the oven and out of the refrigerator, and remember that heat always travels from a warmer to a cooler area.

Iron. Iron (Fe), occurs largely in the form of oxides and carbonates in widely scattered areas of the earth. Because iron is the most abundant and inexpensive of the heavy metals and has so many useful properties, it can be considered quite indispensable to the needs of modern man. For example, iron forms such a large part of the generators that produce our electricity that it may be considered the basic metal of all commercial electricity.

In the United States, the open-pit mines of the Mesabi Range of northern Minnesota have been a major source of iron ore. Within recent years, as former sources of iron have shown signs of depletion, a successful effort has been made to mine taconite, a hard, magnetic material containing iron. The taconite is mined by jet drills; then it is pulverized and the iron present extracted by magnets and formed into pellets ready for the blast furnace. Through the process of smelting the iron ore with carbon and limestone, the iron is freed from its impurities, although some carbon always remains.

PIG IRON. Molten iron from the blast furnace is cast into pig iron (the castings or ingots are called "pigs") and this is the basic metal from which other iron products are made. Pig iron contains 5 to 6 percent carbon and small amounts of manganese, sulfur, silicon, and phosphorus. When the carbon is combined with the iron, white iron, a hard, brittle substance that cannot be machined, is formed. If the carbon is uncombined and remains in flakes throughout the iron, gray iron, a tough product which is easily machined, is the result. Mottled irons are found between white and gray and are used for castings.

MALLEABLE CAST IRON. Malleable cast iron is formed by annealing the iron at a high temperature, a process that causes the surface carbon to be oxidized. Such iron is no longer brittle.

ARMCO IRON. Armco is the trade name for a pure iron produced commercially by the open-hearth method. It is guaranteed to have not more than 0.16 percent of impurities. Of high tensile strength, Armco iron is a good conductor of electricity, is easily welded, is rust resistant, and takes enamel well.

Steel. Pig iron is the initial product in the manufacture of steel. Impurities are removed from the iron, and the steel is cast and then forged or rolled into final form. The various grades of steel depend upon the amount of carbon present: soft steel contains less than 0.1 percent carbon while hard steel contains 1.0 percent of more, and there are a large number of grades between, depending on carbon content. Because soft steel can be welded, it is used for the frames of ranges, refrigerators, washers, and dryers. Hard steel, formed by heating and then quenching immediately in water, is used for razors and knives where a sharp cutting edge is desired.

STAINLESS STEEL. Stainless steel, often known as 18-8 steel, has had some of the carbon replaced by chromium and nickel—18 parts of chromium and 8 parts of nickel. Stainless steels resist corrosion, have high tensile strength, take and hold a high polish, are not scratched by scouring powders, and are not affected by food alkalies and acids. However, they do not conduct or absorb heat very well, and in order to use stainless steel in cooking utensils, the conductivity must be improved by plating copper or aluminum on the bottom surface of the utensil or by placing a core of copper or iron alloy between two sheets of the stainless steel. Stainless steel utensils not so treated may develop hot spots and may require additional liquid and/or very low heat in the cooking process to prevent sticking. Soap and water will clean stainless steel utensils easily, and no special care is need to maintain their attractive appearance.

VANADIUM STEEL. Vanadium is frequently added to steel made into cutlery. It does not tarnish or discolor, and it increases the ability of the steel to take and keep a sharp edge.

GALVANIZED STEEL (OR IRON). Galvanized steel has a coating of zinc over the steel base to protect against rusting. Because zinc is not ductile, the steel product must be shaped before the zinc coating is applied. The galvanized steel is moisture resistant and is used primarily for articles which come in contact with moisture—pails, washtubs, washboards, watering pots, fruit-jar lids, for example.

Galvanized steel, once widely used for kitchen tabletops, has been replaced by better materials because its heavy coatings tended to crack and peel. The zinc coating on articles of galvanized steel may darken with use, but it can be scoured.

TINNED STEEL. Tin is another protective coating for steel. Unlike zinc, tin is pliable, and sheets of steel may be coated with tin before they are made into utensils. Tinware, long popular in both England and the United States, is graded on the quality and weight of the base metal as well as the tin coating. Block tin indicates the most heavily tinned sheets of steel. The coating should be free from imperfections that would allow moisture to reach the steel base and cause corrosion.

Tin is bright and shiny when new, but as it is used it darkens and becomes a good absorber of radiant oven heat. The dark tarnish acts as a protection and should not be scoured. To keep tin utensils in good condition, wash them with soap and water, rinse with hot water, and dry at once, thoroughly. If food sticks to the surface of a utensil, allow it to soak; do not scrape—that might damage the surface and allow moisture to reach the base metal, causing rust to form.

Aluminum. Aluminum (Al) forms approximately 8 percent of the earth's crust, making it the third most abundant element next to silicon and oxygen. Primarily found in clay, it is obtained commercially from the oxide bauxite, largely Al_2O_3. After the oxide is separated from the clay, it is reduced to metallic aluminum by an electrolytic process developed by an American, Charles Martin Hall. The Aluminum Company of America, now Wear-Ever Aluminum, Inc., was formed as a result of Hall's discovery.

Aluminum has many desirable characteristics: it is light in weight—one cubic foot block of aluminum weighs 169 pounds, compared to the weight of an equal volume of steel, 489 pounds; of copper, 557 pounds; and of lead, 710 pounds. It is also strong, nonmagnetic, noncorrosive, ductile, easily polished, and a good conductor of heat and electricity. And it can be drawn, extruded, welded, and brazed. Aluminum is, however, darkened by alkalies in water and food. The film produced may be removed by lemon or vinegar-water or by cooking acid foods such as tomatoes or rhubarb in the utensil.

Aluminum has more than 2,000 uses, and while some of the products may seem rather distantly related to household equipment, they are all a part of our everyday lives. Our homes, both permanent and mobile, may be sheathed in aluminum, and window frames and balconies are made of it. Automobiles, buses, ships, airplanes, railroad cars, highway bridges, fences, signposts, light standards, and high-voltage wires are all made of lightweight, long-lasting aluminum.

In the kitchen, aluminum is used in the refrigerator for the evaporator, ice-cube trays, shelves, and doors. As foil it may be formed into disposable oven liners and reflector-bowl liners for the surface burners. Aluminum cooking utensils have been made for more than seventy years, and these utensils, which are either polished or given a satin or hammered finish, spread the heat over their surfaces, eliminating hot spots. Range manufacturers recommend the use of a medium-gauge aluminum saucepan on thermostatically controlled surface units. Agitators in clothes washers may be made of aluminum also.

Because aluminum is tasteless, odorless, nontoxic, and impervious to greases, fats, oil, moisture, gas, and light, it is used for bottle and jar tops and for soft-drink and other cans and containers with pull-top openings as well as for wrappings for all kinds of food. When used for wrapping frozen food, it shapes to the food, shutting out air that tends to draw moisture from frozen products.

Aluminum oxide, Al_2O_3, is found in powder and granular forms and is used in ceramics, abrasives, toothpastes, and in paints and pigments for house and roof coverings. It is heat, erosion, and moisture resistant. $Al_2O_3 \cdot 3H_2O$ (aluminum oxide trihydrate) finds application in textiles, plastics, varnishes, and toothpastes.

PORCELAIN ENAMEL AND OTHER FINISHES. Large appliances are usually finished in porcelain enamel. This material covers the exterior surfaces of ranges and the interior surfaces of range ovens, refrigerators, washers, and dryers. Porcelain enamel can withstand the rubbing of other materials across its surface. For example, when fabrics are washed and dried automatically, the tumbling of linens and clothes inside the drum causes continual rubbing against the enameled surfaces, but there is no damage.

Porcelain enamel is a glass-like material made of inorganic substances: silicates, carbonates, and borates. These minerals are fused together in an electric furnace until a clear liquid is obtained. This liquid is drawn off into cold water, and because of the sudden contraction, breaks into fine particles known as frit. The frit is then pulverized, mixed with clay and water to form a thick cream, and stored to age.

Before porcelain enamel is applied to a surface the appliance has already been cut and shaped into its permanent form to prevent the enamel from chipping or crazing under strain. After forming, the appliance is cleaned in an acid bath, rinsed, then dried, and given a coating of zinc or Bonderite, a product of the Parker Rust Proof Company, both of which are rust resisting and protective of the steel base. The ground coat of porcelain enamel contains, in addition to its mineral ingredients, cobalt and/or nickel oxides which have an affinity for iron and steel, making a strong bond with the metal. These oxides are also responsible for the enamel's dark blue or black color. Two additional coats will be required to cover the dark color unless titanium oxide is added to the enamel. Titanium oxide, which increases the opaqueness of the enamel so that only one additional cover coat is needed, has a very pure and white tone and has largely replaced other enamels which usually show a noticeable difference in whiteness from appliance to appliance. If colored enamels are desired, metallic oxides are added. The oxides fuse into the enamel and never wear off or fade. Because porcelain enamels can be damaged by acids, they should finally be given an acid-resisting glaze.

Porcelain enamel is applied by spraying the appliance with liquid enamel or by dipping the appliance into it. Then the appliance is run through an enameling furnace where temperatures of 1,400 F to 1,600 F cause the glass to melt and fuse into the pores of the steel. This first frit coat acts as a binding between the metal and the outer one or two coats, which are applied and fused in the electric furnace in a manner similar to the ground coat. In contrast, synthetic enamel, made of organic gums and oils, is baked on at 300 F. Baked enamel can withstand knocks and blows without chipping; therefore, it is used for the exterior surface of refrigerators, washers, dryers, and for kitchen cabinets. It will wear off if subjected to abrasive action or rubbing, and sometimes the baked-enamel finish on cabinet doors wears off where fingers have rubbed against the panel when opening and closing the door. Baked enamel is sold under various trade names, Dulux, for example.

Electroplated Finishes. An electroplated finish is one in which a thin coat of metal, such as chromium, is deposited over the base metal during a process in which the base metal is made the cathode in an electrolytic bath containing chromium salt. Electroplated finishes are used on small electrical appliances—toasters, percolators, and waffle bakers, for example. Chromium, silver in color and easily cleaned by wiping with a damp cloth and polishing with a dry one, is now used almost exclusively in small appliances. Nickel was once a popular metal finish, but nickel has a yellow tinge and tends to acquire difficult-to-remove stains.

Copper and aluminum may be electroplated on the bottoms of stainless steel saucepans to make them better conductors.

ANODIZED ALUMINUM. The anodizing of aluminum is another finish applied by electroplating. During anodizing a film is deposited on an aluminum utensil which has already been either polished or given a satin finish. The utensil—pie plate, tray, bowl, or bun warmer—is made the anode in an electrolytic bath. The anodized finish, shiny or dull depending on the original finish, provides a wear-resistant coating that will not smudge as ordinary aluminum does. Color may be added to the film. While it is fairly permanent and will not wear off with regular use, it may fade if subjected to high oven or dishwasher temperatures. Only the covers of pans should be colored, and they should not be washed in the dishwasher.

Teflon. Teflon, a fluorocarbon resin developed by Du Pont, is another widely used finish for the inside surfaces of saucepans, frypans, baking utensils, and small electrical appliances. Its chief advantage is that foods do not stick to it, making it very easy to clean.

Two types of Teflon have been produced: the first coatings were soft and easily scratched and special tools of plastic, wood, or rubber had to be used to prevent damage to the surface; Teflon II is a hard-base and tougher coating now available for top-of-the-range utensils and certain electrical appliances.

In the original Teflon process, the cookware was first roughened by sandblasting and then sprayed with liquid Teflon. This first coat, which flowed into the crevices and made a close bond with the metal, was baked on at about 450 F. A second coat containing desired color pigments (and twenty different colors were possible) was then sprayed on and the item was baked at 750 F. In the hard-base Teflon process, the cookware is also sandblasted and then a coating of molten metal or ceramic frit is sprayed on and heated to 1000 F. This coating takes the roughened contours of the sandblasted surface and forms an extra hard base, which is covered by the two coats used in the original process. The result is that Teflon II does not have a smooth surface and is therefore not easily damaged; it can be used with metal spoons and spatulas.

Teflon has high heat stability; it is not affected by the usual cooking temperatures, nor is it harmed by food acids and alkalies or by the strongest detergents. Pans with Teflon finishes can be washed in the dishwasher, but it is best to avoid abrasive cleaners.

Since foods do not adhere to Teflon, fats are not usually needed for frying or baking. Persons who are dieting can thus eliminate unwanted fats. Before first use, however, it is a good practice to oil or grease the nonstick surface, especially in the case of griddles and bakeware. Cake and muffin pans used for butter or fruit cakes, or for products with high sugar content, should be greased before each use. Occasionally, the nonstick finish will discolor when subjected to very high heat. You can remove the stain with one of the commercial stain removers designed for Teflon or by a homemade remedy: boil two tablespoons of baking soda mixed with one-half cup of liquid household bleach and one cup of water in the stained pan for five to ten minutes; wash thoroughly, rinse and dry the pan; and wipe the surface with cooking oil.

In spite of rumors to the contrary, extensive testing in both the United States and Europe has proved that no health hazard exists from using Teflon-coated pans for the usual cooking processes. Records show that more than 100 million Teflon-coated items are now used in 25 million different homes in America without ill effect. Teflon is approved by the Food and Drug Administration for "conventional kitchen use," and Du Pont has established its own standards for the application of the coatings. Companies complying with its standards may use the Du Pont certification labels.

FIG. 2.1. Du Pont certification for housewares coated with Teflon nonstick finishes. *(Du Pont)*

Polyimide. Chemically related to nylon, polyimide is used as a coating for the outside of utensils, but not for their bottoms. Application is by spraying on the drawn utensil, which is then baked at about 600 F. Similar to procelain enamel in appearance, it has a shiny surface and is obtainable in a variety of colors. Although polyimide coatings resist damage; are not harmfully affected by water, detergents, or food; and do not peel, they may be harmed by scraping or scratching. The utensils finished with polyimide may be washed in the dishwasher but should not be scoured, and stains from burned foods and grease are removable with car cleaner. Oven utensils of polyimide are especially desirable because they absorb heat and give a crisp crust.

GLASS. Glass in its several forms finds various uses in the home—as window panes in our rooms, goblets on the dining table, vases for our flowers, and utensils for surface and oven cookery in the home. For the latter use, glass must be heat resisting.

Glass is made primarily from silicates, borates, and carbonates fused together and poured into molds. It is then cooled sufficiently slowly for the material to reach an equilibrium. It is usually transparent—an advantage for viewing the cooking process.

Colored glass is obtained by adding metallic oxides during the forming process, and the color is permanent.

Glass utensils used on surface elements need to have water added during cooking because glass is not a good conductor.

Compared to metals, the coefficient of expansion of glass is high, and consequently it cannot be subjected to extreme changes of temperature. Do not run cold water into a hot casserole because the sudden unequal expansion of the inner and outer layers may cause cracking.

Although glass is not a good conductor, it is an excellent absorber of heat. It is thus especially well adapted for oven cookery. When using glass pie plates, cake pans, and casseroles, manufacturers recommend a temperature 25 degrees lower than for metal utensils of the same type. When using a glass saucepan over an electric unit, it is advisable to place a wire grid between the pan and the unit. If the pan is in direct contact with the unit, the heat may cause a slight softening of the bottom surface, making it uneven after it cools.

As noted above, glass may be shattered by sudden changes in temperature, especially from moisture. It can also be broken by a severe blow, or by dropping on a hard surface.

PYROCERAM. Pyroceram is a nonporous ceramic substance developed as a material for the nose cones of rockets to withstand extreme changes in temperature. Corning Glass Works is now manufacturing utensils of pyroceram that can be used in all types of cookery—surface, oven, and broiler. Pyroceram is very resistant to scratching, crazing, and cracking and is not affected by food acids and alkalies. While it is a good absorber of radiant oven heat, it is not a good conductor. Its ability to withstand extreme changes of temperature makes it possible to take a pyroceram dish out of the freezer and place it immediately in a heated oven.

HEAT-PROOF CHINA AND EARTHENWARE. Heat-proof china and earthenware are clay products and they vary as to the quality of clay used and the type of glaze. Opaque, porous when fractured and coarse in texture, these materials are not usually affected by sudden changes in temperature or by food acids and alkalies. If the glaze is broken, however, utensils of these materials tend to absorb fat from foods cooked in them through their porous material and will impart a rancid flavor when used to prepare future foods.

PLASTICS. Plastics have become an indispensable part of the equipment of the home—sometimes by themselves and sometimes as a part of other appliances and utensils. Lightweight and durable, plastic substances are man-made by the action of chemicals, heat, and pressure from such common, everyday materials as coal, petroleum, natural gas, air, water, cellulose, sulfur, salt, and lime.

Plastics are of two types, thermoplastic and thermosetting, depending on their reaction to heat. Thermoplastics are softened by heat, but harden into the original shape when heat is removed; they react this way any number of times without loss of their original physical properties. In contrast, thermosetting plastics are changed chemically by heat and harden into permanent form.

Thermoplastic Products. Products of thermoplastics include vinyl, polyethylene, nylon, polystyrene, and cellulosics. They all have a variety of valuable uses: vinyl for floor, wall, and counter-top coverings; for shower curtains, upholstery materials, and parts of refrigerators and vacuum cleaners; polyethylene for ice-cube trays, mixing bowls, juice containers, squeeze bottles, and food wraps; nylon for bearings in eggbeaters, gears in sewing machines, and refrigerator parts—door gaskets and rollers for sliding shelves; polystyrene for air-conditioning cases, food bins, picnicware, wall tile, lighting fixtures, and the lining and breaker strip of the refrigerator door; cellulose acetates for synthetic yarns and fabrics. All the thermoplastic substances are odorless, tasteless, and nontoxic and are not affected by moisture, fats, or the common household acids and alkalies.

Thermosetting Plastics. The group of thermosetting products includes urea, phenolic plastics, and melamine. Urea is used in electric-outlet panels and plugs, clock cases, cosmetic jar tops, buttons, and picnicware; phenolic plastics, usually dark in color and opaque, in telephones, vacuum cleaner parts, handles on utensils and irons, legs and bases of appliances, and light plugs; and melamine, available in a wide variety of opaque and translucent colors, in dinnerware and mixing bowls. Thermosetting plastics make good electric insulators; are largely unbreakable; and are resistant to heat, moisture, and scratching.

Additional Characteristics and Uses of Plastics. Plastics have such a variety of uses that it is difficult to imagine life without them. For example, in addition to the uses and products already discussed, plastics are used in house paints. These paints flow readily, dry rapidly, resist marring, and when white do not yellow. Plastics coat milk cartons and bread wrappers; and they make some fabrics water repellent. Plastics also bind with other materials to form laminates for knife handles, counter tops, etc.

While thermosetting plastics resist scratching, other types of plastics do not and thus they require special care—washing with lukewarm water and mild soap rather than common household cleaners. Also, some plastics are charred by direct heat while others are affected by high and prolonged heat like that in the oven or dishwasher. Occasionally, certain plastics will break if dropped and others may be shattered by a hard blow.

Polystyrene tile, vinyl film, and melamine dinnerware have been given the Commercial Standard Seal of Approval (CS) by the Society of Plastic Industry with the cooperation of the Department of Commerce. The CS seal covers methods of testing, rating, certifying, and labeling, and it represents an effort on the part of the plastic industry to encourage the manufacture of high-quality products. (Adoption by a manufacturer is optional.)

SILICONES. Silicones, with some of the properties of glass, form a distinctive group between inorganic and organic substances. They find application in a number of household appliances: on the surfaces of ice-cube trays, waffle grids, and certain pans to prevent sticking; as a stainless, odorless rubber in the gaskets of steam irons to prevent leakage; as an oil to improve the accuracy of clocks on range ovens and the timers in toasters; in thermostats of the probe-tube type used on small electric appliances; and as insulation on electric wires and motors.

MATERIALS USED FOR INSULATION. Insulating materials are important in the operation of the home. As already noted, heat always flows from a warmer area to a cooler one, and insulation helps to keep the heat where it was generated (i.e., in the oven of the range) and also keeps warm air out of the refrigerator and freezer where it is not wanted. In the walls of a house, insulation has the double task of keeping the heat in the house during the cold weather—and out of it during summer. In air conditioners and kitchen cabinets insulation decreases the transfer of sound and lessens noise; on the handles and feet of small appliances it prevents burning during handling and damage to the supporting surface. The insulation on electric wires keeps the current channeled and decreases the possibility of a short circuit.

Because of the nature of their uses, insulating materials must be impervious to odors and moisture, and they must also be unaffected by high temperature, burning or charring, and rodents. Also essential is that the materials be light in weight and not sag or deteriorate with the passage of time.

FIBERGLAS. Fiberglas is the registered trade name for certain man-made inorganic fibers, largely silicates, manufactured by the Owens-Corning Fiberglas Corporation. The mineral ingredients of Fiberglas are melted and then blown into cooling chambers or allowed to fall in a spray from the top of a tower. The product that results will depend on the proportion of the ingredients and the temperature used. The diameter of the fibers are made to vary so they may have different uses.

The two general types of fibers are: fibers that form a resilient, fleece-like mass suitable for making into blankets, bats, and boards for oven, refrigerator, and water heater insulation, and for dampening sound; and a second type that is suitable for weaving into tapes, braids, and cloth. Fiberglas is flexible, of high tensile strength, does not absorb moisture, is noncombustible, and will not rot, mildew, or rust.

Fig. 2.2. (a) Types of fiberglas insulation to control heat, cold, and sound. *(Owens-Corning Fiberglas Corporation)* (b) The blanket of Fiberglas is wrapped around oven and broiler to keep the heat inside. *American Gas Association)*

URETHANE FOAM. Another type of insulation widely used in refrigerators and refrigerator-freezer combinations is urethane foam, made by mixing liquid $Freon_{12}$ with other chemicals. In a reaction which takes place in the space between the refrigerator walls, $Freon_{12}$ changes to an expanding foam, filling all the space with millions of Freon cells, and bonds to all the surfaces to prevent the passage of heat. This foam is moisture resistant and 1¼ inches of it is equal to 3 inches of the conventional insulation, allowing an increase in internal area of the refrigerator, while occupying the same floor space as the older model.

ROCK OR MINERAL WOOL. Mineral wool is a stable insulating material which will not burn or deteriorate and does not attract vermin and rodents. Although Fiberglas has largely replaced it as an insulation in household appliances, mineral wool is still used for blankets, bricks, insulating cement, pipe coverings, acoustical plaster, and occasionally in the walls and floors of houses. The method of manufacture is similar to Fiberglas.

ASBESTOS. Asbestos, made from serpentine (hydrous magnesium silicate), occurs as flexible, slender fibers—some long and some short. The long fibers are spun into yarns and woven into cloth; the shorter ones are made into sheets (such a sheet formed into a plate and bound with a metal is occasionally placed under a saucepan to keep the contents from overheating).

Asbestos is heat resistant, noncombustible, and light in weight. It is not, however, moisture proof unless especially treated. Neoprene, a synthetic rubber, has largely replaced the use of asbestos in the cords of small appliances.

MICA. Mica, a silicate, occurs in nature in the form of inch-thick plates. These plates are trimmed and cleaned and split into pieces, 0.002 inch thick. The pieces are then joined into sheets with a binding material. The best grades are very transparent, and while they are flexible and soft, they are also strong. Mica is a poor conductor of electricity, but a fairly good conductor of heat, characteristics that make it useful for small appliances. For example, the nichrome heating element in toasters is frequently wound around mica sheets.

WALL, FLOOR, AND COUNTER COVERINGS

WALL COVERINGS. The most widely used finish for kitchen and bathroom walls is undoubtedly painted plaster. Several coats of high-grade semigloss paint applied to plaster walls produces a surface that is hard, smooth, and impervious to moisture. Since painted ceilings are hard to wash, calcimined ceilings are desirable, and they can be redone when necessary.

The wall above a counter is frequently done in either ceramic or plastic tiles, and both are satisfactory, resistant to food stains and moisture, and easily cleaned with a damp cloth or special wax. Plastic tile is somewhat easier to apply over old construction than the ceramic tile. Tiles of aluminum, stainless steel, and copper have adhesive backs and can be used for "do-it-yourself" installations. Painted or synthetic-enameled wallboard and composition linoleum are also available. The linoleum has a flexible adhesive fabric backing and is resilient, waterproof, and stain resistant. Certain textured wall coverings of plastic in a variety of patterns look like real fabric and can be purchased by the yard.

FLOOR COVERINGS. Floor coverings should be resilient and soft beneath the feet, but not able to be dented by heavy furniture and not easily penetrated by dirt. They must also be resistant to moisture and stains from household oils, acids, alkalies, and they must be able to be cleaned easily.

In addition to carpeting, the Building Research Institute defines six basic materials for floor coverings: vinyl in tile and sheets; linoleum; asphalt tile; rubber and synthetic rubber sheet and tile; and cork. Ceramic floor tile is also used, most frequently in entrance halls and in bathrooms. Its lack of resiliency makes it undesirable as a covering for kitchen floors.

Vinyl Sheet and Tile. Vinyl products are made from vinyl resins, fillers, pigments, and plasticizers. In one case, asbestos fibers are used. These materials, when heated together and pressed into shape, are resilient, resistant to moisture, stains, and indentation, and are long-wearing. In fact, wear tests on floor coverings made by the Agricultural Engineering Department of the Texas Agricultural Experiment Station found that sheet vinyls and rubber tile showed the least wear. They graded the six as follows: least wear, sheet vinyls and rubber tile; most wear, asphalt tile; cork, linoleum, and asbestos ranked between. Dry abrasives caused less wear than wet abrasives, and waxing on cork and linoleum increased resistance to wear. (B. R. Stewart, O. R. Kunze, and Price Hobgood, "Accelerated Wear Tests on Common Floor-covering Materials," *Texas Agr. Exp. Sta. Bul.* 890, College Station, Texas, March, 1958.)

Linoleum. Linoleum is manufactured in at least five types—plain, marbleized, Jaspe, inlaid, and printed—and in several thicknesses and a wide variety of designs and colors. Plain linoleum is one-color; marbleized linoleum looks like a sheet of marble, is available in various colors and does not show soil; Jaspe has a striated appearance and is two-colored. A thick battleship linoleum was so named because of its use on United States battleships.

The colors in Jaspe and marbleized linoleum go through to the burlap back. Inlaid linoleum is made by placing plain, marbleized, or Jaspe on burlap and the colors also go through to the burlap. Sometimes the designs of this type of linoleum are raised making it attractive but more difficult to clean. In printed linoleum the design is printed on the surface in oil paint. It is also attractive, but it is less expensive than other types, and the pattern wears off when subjected to heavy usage.

For long, satisfactory service, linoleum should have an underlay of felt or sponge rubber. The underlay, cemented to the floor, prevents cracking and buckling, covers any rough or uneven area, and tends to make the floor more resilient. The floor will also be warmer in cold weather and quieter. Some linoleums are laid without an underlay. Printed linoleum is often placed directly on the floor, but it will wear longer if over an underlay. And felt-base linoleum can be cemented directly to the floor. The seams of the underlay are joined with waterproof cement.

Asphalt Tile. Asphalt tile, obtainable in 9 by 9-inch tiles, is made from resin hydrocarbons and asbestos fibers with color pigments and inert filler added. Colors go through to the back. The tiles are pressed into shape when hot. Asphalt is resilient and is the first such floor covering that can be laid on top of concrete. It resists moisture and fire and is quiet, odorless, nonabsorbent, and economical. The tiles can be damaged by grease and oil (new varieties are grease-proof) and can be dented by heavy furniture.

Ceramic Floor Tile. Used to some extent in bathrooms, ceramic floor tile is also used occasionally in kitchens. It is waterproof, but it is not resilient, making it difficult to stand on for even short periods. Mediterranean floor tile, long used in Italy and southern Europe, is now used in this country as well because it is hard, attractive, and excellent for use on patios and terraces.

Synthetic Rubber. Since 1946 all rubber floor coverings have been synthetic in nature, made from styrene-butadine rubber, clay, oils, resin, waxes, whiting, sulfur, color pigments, and other ingredients. Sulfur vulcanizes the rubber, helping it to maintain its shape under conditions of stress. Patterns are marbleized in black and white and various other colors.

Floors of synthetic rubber are quiet, resilient, resistant to indentation and wear. They may be washed with water and a mild detergent and waxed, preferably with a water-emulsion wax. Avoid all cleaning materials with solvents, oils, lacquers, shellacs, and varnishes.

Cork. Cork tiles are made from granules from the bark of certain oak trees, bound together with phenolic or vinyl resins, and pressed into shape under moderate heat. Such tiles are resilient and have high tensile strength. Since they are wood product, they can be damaged by moisture and swell when wet. Vinyl resin tends to form a protective coating over the cork. Before installation the tiles must be conditioned to room temperature for at least 24 hours.

Carpeting. Carpet material of nylon, polypropylene, and acrylic backed by foam rubber is now available for the kitchen, bathroom, laundry, and even the furnace room. The carpet, purchased in a roll or 12-inch tiles, is held in place by the suction action of the rubber backing. It may thus be removed in whole or in part to be cleaned. If soiled areas are neglected, they become a source of bacteria and result in unsanitary conditions and foul odors. For this reason there have been some differences of opinion on the desirability of carpeting in the kitchen and bath.

Maintenance Requirements of Floor Covering Materials. Excessive water and strong and abrasive cleaning materials should be avoided in the maintenance of floor covering materials, most of which are somewhat porous.

Water-emulsion waxes may be used on all floor coverings except cork. They form a protective finish by filling the pores. Best results can be achieved if two coats of wax are applied, the second after the first has thoroughly dried, and if the wax is renewed from time to time. If dirt penetrates the wax and builds up in areas where the floor is not travelled over, i.e., behind doors, under tables, etc., the wax will have to be removed by a commercial wax remover or household ammonia (always use in a well-ventilated room and give some protection to the hands). Whatever wax is used, it must not be slippery. Water-emulsion waxes are not generally slippery, and if applied correctly in a very thin coat and thoroughly buffed, neither are paste waxes.

Certain products are specifically manufactured for cleaning particular floor coverings. The Asphalt and Vinyl Asbestos Institute and the Rubber Manufacturers Association have lists of approved cleaners for their products. For general information on all floor maintenance, Publication 597, *Installation and Maintenance of Resilient Smooth-Surface Flooring,* Building Research Institute, National Academy of Sciences–National Research Council, 1958, is extremely valuable.

COUNTER-TOP MATERIALS. A number of materials–linoleum, vinyl, and ceramic tile, for example–used on the floor may also be used for counter-top finishes. Other finishes include stainless steel, aluminum, and such plastics as Formica, Micarta, Textolite, and other plastic laminates.

Chopping blocks of wood are often installed as a section of the counter, preferably near the range. Cutting on a wood surface prevents damage to the counter-top material and the cutting knife. Beech, birch, and hard maple can be used, but maple is least affected by grease. The wood countertop should be sealed to resist moisture and stains and should not be subjected to temperatures higher than 350 F.

Linoleum and plastics are widely used for counter tops. They are durable, resilient, quiet, resistant to moisture and most household stains, and are obtainable in a variety of attractive colors and patterns. Waxing helps to preserve the surface but they can be damaged by high temperature and abrasive action.

Laminated counter tops are made from several layers of special-type papers bonded together with synthetic resins under pressure and heat. The surface layer may be of Melamine resin, a very heat-resistant substance. Laminates are rigid and tough, nonporous, and easily cleaned. The counter edge may be of the same material as the counter, a metal band, or a combination of the two.

Stainless steel and aluminum, frequently anodized, are counter-top materials resistant to moisture and stains but having several drawbacks. Stainless steel will show waterspots if not wiped dry when washed, and it can also be marked by a knife. Both these materials will conduct electricity and care must be taken when operating small electrical appliances. These materials should be installed with an insulating strip between the counter top and the sink. Unless insulation is placed under these materials, they may be noisy, but not as noisy as ceramic tile.

Ceramic tile is moisture- and heat-resistant, but it is not as desirable for a counter top as most of the other materials discussed. It is noisy, has crevices where dirt can collect, and dishes are easily damaged by its hard surface.

In general, a counter-top material should be chosen because it is quiet, unaffected by common household stains and heat—both dry and moist—and resistant to abrasives. If these guidelines are followed, the material will give long service for the money expended.

PROJECTS AND EXPERIMENTS

1. Examine the materials your instructor has assembled on the laboratory table. Identify the basic materials and finishes. Note the features that contribute to or hinder ease of cleaning.
2. Following recommended instructions, clean an item of porcelain enamel, aluminum, and stainless steel. Also clean a Teflon-coated utensil, the door of a laboratory cabinet, and a counter top.
3. Determine the thickness of the insulation of the refrigerators and ranges in the laboratory. Does the thickness of the insulation give a clue as to type?
4. Test the efficiency of the oven insulation of one gas and one electric range. Heat each oven for one hour at 400 F. Determine any hot spots on the top, sides, and door areas by placing a thermometer bulb on the surfaces.
5. Examine the laboratory floor covering for discoloration and abrasion. Also examine the kitchen floor where you live. Report the results to the class and make recommendations for the type of care needed.

3 LAUNDRY AIDS

CHARACTERISTICS OF WATER. Water is spoken of as hard or soft—terms which refer to the presence or absence of certain insoluble salts. Soft water is by far the more desirable for use in laundering clothes; unfortunately, however, hard water is much more prevalent in the United States with 80 percent of the states having water of various degrees of hardness. Soft water is found naturally only in Washington and Oregon and in the Atlantic Coastal and Gulf states.

HARD WATER. When rainwater seeps through the soil, it picks up soluble inorganic salts. Water is permanently hard when it contains calcium and magnesium sulfate and chloride. Calcium carbonate (limestone) is insoluble in water, but rainwater usually contains carbon dioxide (CO_2) in solution, and the carbon dioxide reacts with the calcium carbonate to form soluble acid carbonate, $Ca(HCO_3)_2$:

$$CO_2 + H_2O + CaCO_3 = Ca(HCO_3)_2$$

Calcium acid carbonate causes temporary hardness. When the acid carbonate is heated, the reaction is reversed:

$$Ca(HCO_3)_2 \text{ heated} = CaCO_3 + CO_2 + H_2O$$

and the water is softened. The calcium carbonate will appear as a scaly deposit in the water heater and pipes (Fig. 3.1). Notice this deposit when heating hard water in a teakettle or saucepan. Permanently hard water, containing calcium and magnesium sulfate and chloride, cannot be softened by heating and appropriate chemical methods must be used to correct its undesirability for getting clothes properly clean.

FIG. 3.1. This deposit of solidified scale in a pipe through which hot water flowed occurred in less than two years.

Degrees of Hardness. Water is classified into the following degrees of hardness:

Grains of hard water salts per U.S. gallon	Degree of hardness
0 to 3	soft
3 to 6	moderately hard
6 to 12	hard
12 to 30	very hard

Note that measurement is by weight. It is sometimes expressed as parts per million (ppm). One grain per gallon is equal to 17.1 ppm.

SOFTENERS

Zeolite. It is frequently desirable to soften the water system of the home, and modern central softening systems use zeolite, a complex chemical of sodium aluminum silicate ($2SiO_2$, Al_2O_3, Na_2O), and/or a bed of synthetic resin beads in a tank (something like a small water heater), containing a bed of gravel or other filtering agent.

As the hard water filters through the zeolite, ion exchange takes place and the sodium replaces the calcium and magnesium, which form calcium and magnesium zeolite. When the sodium in the zeolite has been wholly replaced, the zeolite must be regenerated by adding common salt to the tank. The sodium in the salt takes the place of the calcium and magnesium which are flushed out. Automatic regeneration of the tank is possible by placing a tank for dry salt beside or in the zeolite tank. An electrically controlled time switch will cause the salt to be added as needed. Water softening service is also available from commercial companies, and the tank is replaced at regular intervals.

If both cold and hot systems cannot be softened, the hot system should be the choice since we have seen that heating hard water tends to cause deposits in heaters and pipes. Some persons however prefer unsoftened water for drinking, and certain investigations report that hardening of the arteries is more prevalent among people who drink soft water. Some cities reduce water hardness, and remove other undesirable aspects of the water supply such as iron and turbidity, and finally chlorinate the water to destroy any bacterial content.

Unit Softeners. When the water is hard and the water system of the house is not softened, some type of softener must be used with each washer load.

WASHING SODA. Homemakers in hard-water regions originally used washing soda (Na_2CO_3) as a water softening agent. For example,

$$CaSO_4 + Na_2CO_3 = CaCO_3 + Na_2SO_4$$

All sodium salts are soluble, but the calcium carbonate formed an undesirable precipitate that floated on the water as scum or settled to the bottom of the washer tub.

CALGON. Calgon is representative of a group of nonprecipitating water softeners that includes Noctil, Spring Rain, New Oakite, and Miracle White. Calgon as a home laundry aid grew out of research done to eliminate deposits of calcium carbonate in boiler pipes. It was found that sodium metaphosphate when used in hard water prevented the precipitates of calcium and magnesium carbonates from forming, because the calcium and magnesium ions were slightly dissociated in the water (a process known as sequestering).

Moreover soap did not react with the calcium and magnesium to form a curd.

Calgon, a trade name for sodium metaphosphate, was originally tested by the Household Equipment Department of Iowa State University. After numerous experiments using hard water, it was found that Calgon used in the first rinse water would remove any detergent residue left in the fabric and that clothes with a buildup of lime soap and a grayish tinge would be reconditioned if a cup or more of Calgon were used in the washer with no other laundry aid, the process being repeated until suds no longer formed.

Calgon, of course, is a phosphate, and future studies will tell whether its use is to be continued or whether some other means will be found for softening water.

OTHER UNDESIRABLE CHARACTERISTICS OF HARD WATER. Hard water is occasionally high in other characteristics that adversely affect laundering. These various factors tend to cause yellow and brown stains on fabrics, manganese stains from certain dyes, iron from ground water or rusty water pipes.

As small a concentration as 0.1 ppm of iron can cause poor color in table and bed linens, for example. Iron in city water is usually in the ferrous form and is soluble and invisible. When it comes in contact with the air, however, it is oxidized to ferric iron and precipitates out as a yellowish product. Heating hastens the change and the use of chlorine bleach accelerates the process. Consequently, care should be taken to avoid using chlorine bleach in water containing iron. Iron may be held in suspension in the wash cycle by water softeners and detergents, but may settle out during the rinse cycle. Subsequent ironing makes the discoloration more visible.

Turbidity. Finely divided organic substances and silt and clay produce turbidity in the water and will sometimes cause graying of the clothes.

Alkalinity. Water high in carbonates tends to be alkaline and may cause cottons to yellow in the dryer or during ironing. The pH scale has been adopted as an easy method of indicating the acidity or alkalinity of a solution. The scale is numbered from 1 to 14 with number 7 the neutral point. Numbers 8 to 14 indicate alkaline conditions, numbers 1 to 6 acid. The numbers near 7 show mild conditions, that is, No. 8 is mildly alkaline, No. 5 mildly acidic. The end numbers give strong reactions; for example, No. 13 would be very alkaline, a No. 2 solution, very acidic. Most soaps and synthetic detergents give most satisfactory results in an alkaline solution between 10 and 11. It should be remembered that pH is a logarithmic scale and each change of one pH unit means a ten-fold change in acidity or alkalinity.

DETERGENTS. Detergent is a general term that includes both soap and synthetics and means a substance used as a cleaning agent.

Water by itself is not efficient in removing soil. Water alone has high surface tension, the force that will cause any liquid to take the form of spheres or drops, a form that gives the smallest surface area in proportion to volume, as if a film were stretched tightly around the waterdrops. A detergent is needed to lower the surface tension of the water, causing it to spread out in a film (instead of staying in drops) and wet the fabric being washed.

Soil is then separated from the fabric by use of water and a detergent which holds the soil in suspension until it can be removed by agitation, extraction, and rinsing. Inorganic soil—dust and mud—is removed by adsorption; the soil clings to the surface of the foam film. Organic soils—fats and oils—are removed by emulsification; the soap bubbles surround the greasy materials until they are agitated away. Other organic soils, fruit and vegetable stains, blood stains, other food spots, body excretions, and epithelial cells mixed with dirt often need additional treatment before the washing process.

SOAPS. Soaps are made by treating certain oils and fats with lye (NaOH). For example, the fat stearin, $C_3H_5\ (C_{17}H_{35}COO)_3 + 3NaOH = 3C_{17}H_{35}COONa$, which is soap. In the reaction, glycerine, $C_3H_5\ (OH)_3$, is also formed. Other fats, olein, palmitin, and tallow are often used, together or alone. Soap then is the sodium salt of a fatty acid and in its manufacture glycerine is a by-product.

Soaps are divided into two groups, mild and built. Mild soaps are pure with no free alkali and are used for washing delicate fabrics, wools and silks, and certain man-made fibers. Built soaps have some free alkali left from the manufacturing process or added later, or other cleaners may be added such as silicates, carbonates, phosphates, or naphtha. Even a pure soap when dissolved in water causes the formation of a little alkali:

$$C_{17}H_{35}COONa + H_2O = C_{17}H_{35}COOH + NaOH$$

The organic acid ($C_{17}H_{35}COOH$) is weaker in reaction than the inorganic alkali (NaOH).

SYNTHETIC DETERGENTS. Synthetic detergents are used almost universally because of their easy solubility in water of all temperatures and because in general they form no precipitates in hard water. One glance at the detergent section of a grocery store convinces the homemaker that the variety of synthetic detergents far outnumbers the soaps, and in fact, they have replaced 80 percent of soap use in the United States.

Synthetic detergents are sulfated fatty alcohols and esters. They contain the sulfonic group derived from sulfuric acid, a material that can be produced in tremendous quantities, reacts easily with a variety of other substances, is inexpensive, and combines readily with petroleum and benzene which are widely used in synthetics. Chemically, it is formulated:

$$CH_3(CH_2)_mCH_2SO_3H$$

When the hydrogen of the sulfonic group (SO_3H) is replaced with sodium, we obtain the sodium salt of the sulfonic compound, a synthetic detergent.

Detergents are colloidal electrolytes, as are soaps. A colloid consists of very fine divided particles of a substance, often ultramicroscopic in size, dispersed in another medium (fog: water particles dispersed in air; milk: fat particles dispersed in water). An electrolyte is a substance in a liquid state that dissociates into free ions, and so is capable of conducting electricity. The detergent is made of two parts, the hydrocarbon chain of various lengths, depending on the number of carbon atoms, and the sodium atom.

Detergents are classified as: light (pure) detergents and heavy-duty detergents. The latter contain "builders" that give superior cleaning performance at a lower cost than pure detergents. The builders are the same in both detergents and soaps: silicates, borates, phosphates. They neutralize acid soils, dissolve water insolubles, emulsify soil particles, react with grease and soils to form soaps and therefore make them soluble in water, and help to maintain the pH of the soil. Heavy-duty detergents may be high- or low-sudsing, the low-sudsing type is recommended for use in a tumbler type washer. Foam is not essential for cleaning, but many homemakers have been conditioned to take it as an indication that the detergent is performing actively.

A list of common detergents, soaps and synthetics, is given in Table 3.1. This table is not all inclusive; certain products are more widely sold in one area of the country, others in another area, and new products frequently appear. A detergent selection should be

TABLE 3.1. Adapted from the table "Let Our Waters Live"

DETERGENTS	Amt. per Washload	Units of Phosphates per Washload
Add-it	½c	0
Culligan Soap	any	0
Diaper Sweet	any	0
Ivory Flakes	any	0
Diaper Pure	1¼ Tbsp	1
Trend	1½c	6
Special-T	½c	7
Instant Fels	1½c	8
Blue Magic	1¼c	19
Amway SA-8	¼c	20
Bestline B-7	¼c	20
Wisk	½c	21
Montgomery Ward	⅔c	21
Surf	1¼c	24
Laundry Maid	1c	25
Gain	1¼c	27
Dreft	1½c	27
Silver Dust	2c	28
Cold Power	1¼c	29
Bold	1¼c	29
Ajax	1¼	31
Cold Water All (liquid)	½c	31
Rinso	1¼c	32

ENZYME PRESOAKS		
Brion	½c	30
Axion	½c	34
Biz	½c	37
Sears	½c	55

ADDITIVES		
Fels Naptha Bar	any	0
Borateem	any	0
Borax	any	0
Right Fabric Softener	any	0
Sal Soda	any	0
Calgon	½c	57

AUTOMATIC DISHWASHING COMPOUNDS (per washload)		
Special-T	1 Tbsp	1
Calgonite	2 Tbsp	6
Advance	2 Tbsp	6
Finish	1½ Tbsp	6
Electra-Sol	2 Tbsp	6
Dishwasher All	2 Tbsp	9
Cascade	2½ Tbsp	11
Amway	2 Tbsp	11

DETERGENTS	Amt. per Washload	Units of Phosphates per Washload
Easy Life Heavy Duty	1½c	32
Cheer	1¼c	33
Fab	1½c	34
Oxydol Plus	1¼c	34
Punch	1¼c	35
Breeze	2c	36
222	¾c	37
Concentrate All	1c	38
Sears	½c	38
Ad	1c	38
Easy Life Enzyme	1½c	38
Duz	1½c	39
Easy Life Blue	1½c	39
Tide XK	1¼c	40
American Family	1¼c	40
Drive	1¼c	41
Cold Water All	1¼c	42
Vim	4 tablets	44
Fluffy All	1½c	52
Bonus	2c	55
Salvo	2 tablets	59
Dash	1c	60

BLEACHES AND BLUING		
LaFrance Bluing	½c	0
Miracle White Bleach	½c	14
Action	1 packet	24
Snowy	¾c	27
Beads O' Bleach	2 oz	41

BOOSTERS		
Climalene	¼c	14
Easy White	½c	24
Anything Goes	½c	27
Miracle White	½c	41

DISHWASHING LIQUIDS
(Generally contain no phosphates)

ALL PURPOSE CLEANERS		
Amway L.O.C.	any	0
20 Mule Team Household	any	0
Ajax Floor and Wall	2 Tbsp	4
Janitor in a Drum	1c	9
Soilax	3 Tbsp	12
Spic and Span	½c	23

Source: National Wildlife Federation, *Conservation News*, July 1, 1970.
Whichever detergent you use, avoid waste, use no more than the quantity specified on the box.
Analyses performed by Northwestern Students for a Better Environment, Cresap Lab, Northwestern University, Evanston, Ill.

based on the type of fabric and fiber, the amount and kind of soil, the composition of water available, and the type of washer. These same points will determine the amount of detergent to be used. In general, homemakers are urged to use less detergent in order to decrease the phosphate contamination of streams, but too small an amount will tend to cause graying and yellowing of fabrics.

DETERGENTS AND PHOSPHATE CONTENT. The extensive use of synthetic detergents has caused the problem of foam on rivers; it has even appeared in drinking water at times. The foam results from the failure of bacteria to break down the chemicals used in manufacturing the detergents, most of which contained the benzene ring, e.g.,

The need to reduce the appearance of foam in sewage plants and in streams has led manufacturers to change the composition of detergents, eliminating the branched carbon chains and benzene types, and producing what are known as biodegradable detergents that can be digested by bacteria. Changing the composition of detergents has been an expensive proposition, but it is gradually taking place. Many of the present detergents are biodegradable. It is part of the effort against pollution.

The study of phosphates and their effect on environment has received much study by several branches of the federal government concerned with public health and the environment, and on September 15, 1971, they joined forces to suggest that homemakers continue the use of detergents containing phosphates. The general opinion was that the ill effects of detergent phosphates had been greatly exaggerated, since only about 13 percent of phosphates reaching rivers and lakes comes from detergents, the other 87 percent results from fertilizer seepage, human and animal wastes, and the runoff from some manufacturing plants. Moreover, some of the materials used as substitutes for phosphates are not an improvement for they frequently do not get the clothes as clean as phosphates do, allowing redeposition of soil on the fabric. They are alkaline and therefore dangerous to keep around children. They cause corrosion in parts of the washer, sometimes even roughening the surface of the agitator fins and this tends to damage clothes fibers.

COLD WATER DETERGENTS. Certain detergents on the market are primarily for use in cold water. A research project was carried on in the equipment laboratories of Iowa State University, in which sheets, soiled by normal use in homes, were laundered for 24 weeks in cold water and also in hot water, with detergents designed for use at the different temperatures. Objective tests by measuring instruments and subjective results from viewing by a homemaker panel indicated no significant differences. The homemakers were not sensitive to variations in whiteness. A consumer panel of college students did note that the laundered sheets had a greater whiteness than the standard unlaundered sheet, apparently the effect of the optical brighteners in the detergents. Other investigators have obtained similar results.

If the only concern of the homemaker is to have her laundry "look cleaner," then a high level of optical brightener will be the most important ingredient in the synthetic detergent, but if she is truly concerned about the amount of soil remaining in the fibers and fabrics in the laundry, she may question whether the optical brightener is masking the residual soil. For those interested in the manufacture and care of fibers and fabrics, this remains a question to be answered.[1]

ENZYME DETERGENTS. Certain detergents have enzymes added to aid in cleaning. This is a fairly new trend. The first detergent containing enzymes was Biotex, which was marketed in Holland in 1963. It soon reached 20 percent of all detergent sales. Noting this success, detergent manufacturers in the United States began importing European proteolytic enzymes and developing them to tolerate the high pH and high temperatures used in our laundry processes, although one Dutch product, Alcalase, a protease enzyme, was alkaline resistant and stable to 140 F. Previously in the United States enzyme cleaning products had been used for many years by dry cleaners to aid in spot removal.

Enzymes are made by microbiological fermentation, using *Bacillus subtilis,* a harmless organism commonly found in air and soil. Different enzymes produce different reactions: protease digests protein substances; amylase, starch; and lipase, fats. *Bacillus subtilis* produces both amylase and protease. Enzymes hydrolize all stains except tea, coffee, and fruit juices, breaking them down to water soluble products. By themselves, enzymes cause only a little cleaning. They work best when combined with the proper detergent or when followed by a detergent wash. They must be compatible with other elements found in detergents; optical brighteners,

1. Mary Kathleen Sullivan, "Comparison of Specified Properties of Cold Water Detergents with Those Designed for Use in Hot Water: Soil Removal and Fabric Deterioration." M.S. thesis, Iowa State University, 1967.

phosphates, and CMC (an antideposition agent, sodium carboxymethyl cellulose). They cannot be used with chlorine bleach, but may be used with perborate. Too acid or too alkaline a solution tends to inactivate the enzyme; it cannot be used with ammonia. Enzymes do not act on nylon, polyester, wool, and other fibers.

When used in a detergent, the enzyme forms from 5½ to 9½ percent of the total. Enzymes are of two kinds: one is primarily for presoak (Axion, Biz); the other is combined with conventional detergent use (Gain, for example). It is suggested that the presoak period be at least one-half hour long and an overnight soak is strongly recommended for stubborn stains and heavily soiled articles. Experiments showed that soaking made sheets and pillowcases whiter and improved the appearance of dish towels and potholders. A presoak often reduces the amount of detergent needed for the wash period.

At present enzymes are used in dry detergents, and they remain stable at room temperatures for months. Usually in solution they lose their activity. Manufacturers are researching their use in liquid detergents. Research is also being carried on in colleges and universities to determine enzyme stability when combined with other substances.

In addition to enzymes, synthetic detergents frequently contain other products that increase their usefulness. Water softeners and builders have already been mentioned, others may include fabric softeners, antibacterial agents, fluorescent brighteners, and a surfactant—a surface active agent that brings the detergent in close contact with the fabric and the soil, and makes the soil more soluble; also a suds stabilizer, and an antiredeposition agent that prevents soil from redepositing on material (CMC). Some detergents also contain corrosion inhibitors to protect the metal of the washer.

If the detergent is to absorb the soil, the hydrocarbon chain must be as long as possible. The temperature selected for the wash solution also depends on the length of the carbon chain. Carbon chains of 10 to 14 carbon atoms form products to be used at relatively low temperatures for the laundering of delicate fabrics. Detergents with 16 to 18 carbon atoms are usually built products and are more effective at higher temperatures. In general most detergents give best results at temperatures between 100 F and 140 F. Mild detergents usually contain sodium sulfate, the built detergents, sodium silicates and/or sodium phosphates. The polyphosphates have been widely used because of their low cost and superior cleaning ability.

Detergents are marketed in the form of powders, flakes, beads, and liquids. With the enactment of the Fair Packaging and Labeling Act of 1966, the U.S. Commerce Department met with representatives of the Soap and Detergent Association to determine the number of package sizes that were available to homemakers.

As the result of a survey of 72 large supermarkets in 72 cities throughout the country, it was found that 29 different package weights of heavy-duty detergents were marketed. Apparently the situation needed improvement. Within a few months a voluntary package simplification program was initiated by the manufacturers. They reduced the package weights to ten. The number has now been further reduced to five, and 95 percent of the total of heavy-duty dry detergents to three weights: regular size, 20 oz; giant size, 49 oz; king size, 84 oz.

WHITENERS AND STAIN REMOVERS. In earlier days, bleaching was obtained by wetting fabrics and allowing them to lie in the sun, repeating the operation as many times as necessary. Sometimes sour milk or buttermilk was used; then lime was used until it was forbidden by official edict. Lime had been used before, as early as 300 B.C., according to some authorities. Modern methods began about 1800.

In removing stains from fabrics, water of one temperature or another should be tried first. Hot water is effective on red, blue, and purple fruit and vegetable stains, and on tea and coffee stains. Cool water and a detergent should be used on greasy stains and on such protein stains as blood and milk. Hot water will tend to set these stains.

If water does not remove the stain, a chemical process of reducing or oxidizing the color to a colorless product must be tried. This process is known as bleaching. However, bleaching usually is not necessary if clothes are always washed in hot, soft water. Textile manufacturers are distressed at the modern tendency of homemakers to use bleaches so frequently. The addition of a bleach dispenser to many automatic washers has apparently increased the use of bleaches.

CHLORINE BLEACHES. Two varieties of bleaches are commonly used in the home, chlorine and two oxygen producing types, hydrogen peroxide and sodium perborate.

Chlorine bleaches are of two kinds: sodium hypochlorite ($NaClO$) in liquid form and calcium hypochlorite $Ca(ClO)_2$, a powder. Sodium hypochlorite is known chemically as Javelle water and is sold under many different commercial names. It is an effective oxidizing agent.

The sodium hypochlorite bleach is used on white linens and cottons, but the calcium hypochlorite bleach may also be used on some man-made fibers: nylon, dacron, orlon, and on mixtures of these fibers with cotton. These bleaches should be used in hot water with a detergent that buffers the action and so protects the fabric, and in the soak or wash cycle to allow for thorough rinsing afterwards, to prevent deterioration of the fibers.

Sodium and calcium perborates, $NaBO_3$ and $Ca(BO_3)_2$, are powders and safe to use on any textile: wool, rayon, nylon, colorfast fabrics, even on fabrics with crease-resistant and water-repellent finishes. The bleaching effect is caused by oxygen set free from the perborate. Hydrogen peroxide, H_2O_2, decomposes readily to form water and oxygen, the oxygen removing the stain. It is used for bleaching silks and wools, and is most effective when used as a fresh sample; it tends to deteriorate with time.

Oxone, a Du Pont product, is a bleach ingredient. It is monopersulfate, a colorless compound, safe for use on all fibers and colorfast fabrics. Beads O' Bleach contains oxone. As with all bleaches, rinse thoroughly after use.

The United States Government publishes a bulletin on all types of stain removal products.[2]

OPTICAL BRIGHTENERS. Homemakers' demand for a white wash has led to the inclusion of increased amounts of optical brighteners in detergents. The first detergents, if they contained brighteners at all, had only 0.01 percent; today heavy-duty detergents contain from 0.1 percent to 0.45 percent. According to some investigators this large amount tends to make fabrics look white, even though some soil remains on the fabric. Most white textiles have a slightly yellow tinge. The blue light of the fluorescent brightener is complementary to this yellow color and the mixing of the two colors produces white. Even the whiteness produced by a bleach can be improved by the subsequent use of a brightener. A larger percentage of brightener is absorbed from a built detergent than from an unbuilt one. Once the brightener is firmly fastened to a fabric, it is not affected by a bleach; consequently the bleach should not be added until sufficient time has been allowed for the

2. "Removing Stains from Fabrics, Home Methods," *Home and Garden Bul.* 62, fifteen cents. Send to Superintendent of Documents, U.S. Government Printing Office, Washington, D.C. 20250.

brightener to become attached. Some brighteners are affected by sunlight and may be partially destroyed if clothes are line dried. Some detergents contain more than one type of brightener so that fabrics of blended fibers may be evenly whitened.

Bluing is used less frequently than formerly because so many detergents contain fluorescent dyes that give the same white effect. Bluing in flake or bead form is added to the wash cycle; in liquid form, to the rinse.

FABRIC SOFTENERS AND ANTISTATIC PRODUCTS. To prevent the tendency for certain man-made fibers to develop static electricity and cling, antistatic products are available for use in the rinse cycle. The buildup of static electricity is caused by a lack of moisture in the artificial fibers. Vegetable fibers (cotton and linen), and especially animal fibers (silk and wool), contain a certain amount of natural moisture that eliminates this buildup by supplying a path over which the electricity can travel and dissipate. The fabric softener usually also prevents the tendency to cling.

The fabric softener is a dialkyl quaternary ammonium salt in a solution of water and alcohol. Perfumes and dyes may be added for pleasing effects. The surfaces of the fibers are coated with a fatty lubricant, resulting in a soft, wrinkle-free hand. The softener also prevents matting of knits, wools and pile fabrics; reduces abrasion, thus increasing tear strength; aids in preventing redeposition of soil; and may shorten drying time. Resin finishes are not damaged by the softener. However, the softener does not work with synthetic detergents, and should always be added in the last rinse after any water softener and detergent have been removed; otherwise, the combination of detergent and softener may cause a dark stain on the fabric. Exceptions to this rule are certain recently marketed combinations of detergent and fabric softeners that can be used in the wash cycle.

If too much softener is used repeatedly, wash period after wash period, the fabric tends to becomes water-repellent, a condition particularly undesirable in linen napkins, bath towels, and a baby's diapers. To avoid this problem, use the softener less frequently, perhaps only every third or fourth wash period. Sometimes a softener causes yellowing, but it will soften all types of fabrics, even heavy denim, making ironing easier and faster, making woolens fluffy, and minimizing wrinkles.

If bluing is used, use the softener after the bluing rinse; use the softener in the rinse water; never pour directly on the material.

STARCHES. Starches are no longer widely used, but a few men still like the collars and cuffs of their shirts starched, and starch adds charm to a small child's bright-colored cotton dress and perhaps to the kitchen apron. Starch increases the life of garments, it sheds dirt and dust and keeps many fabrics clean longer than if unstarched. It is a sizing and it restores the smooth finish that may have been washed out of a fabric. It tends to prevent excessive wrinkling.

Starches are vegetable or plastic in composition. Vegetable starches are sold in powdered, lump, or cube forms, and prepared with either boiling or cold water or sold as a liquid, which may be diluted to any desired strength. Plastic starches are made from resins and need be used only every third or fourth wash period. A soluble synthetic starch, a CMC (sodium carboxymethyl cellulose) agent that can be washed out during the next laundry process, but gives fabric body without stiffness, is marketed under the name Glosstex. Vegetable starches coat the surface of the fabric, plastic starches penetrate the fibers. When vegetable starches are used with dark materials, the starch solution should be colored with bluing, tea, or a dye, to prevent a streaked appearance on the garment.

Liquid plastic spray starches in aerosol containers are available for use during ironing and are used to starch small areas. The stiffness can be controlled by the amount of spray used. Spray starches may contain perfume, and usually a stabilizer, an ingredient to increase shelf life without drying in the container.

PROJECTS AND EXPERIMENTS

1. Determine wetting ability of water. Fill each of two custard cups half-full of distilled water. To one cup add three drops of liquid tincture of green soap (may be purchased from drug store). Using separate medicine droppers place a drop of the clear water on a 1-inch square of wool flannel, and a drop of the soapy water on a second square of the wool material. Float additional squares of the wool on the water in each custard cup. What causes difference in action? What principle is involved?

2. Compare characteristics of hard and soft water. Place one cup of distilled water in clean bottle or fruit jar, one cup of hard water in second container. To each add 20 drops of green soap solution. Shake. Note appearance of contents and difference in sound when solutions are shaken.

3. Visit a local grocery store. Count the various detergents displayed, noting sizes available. Determine value of information found on

packages. How many contain phosphates? Are any advertised as being free from phosphate?

4. Arrange a visit to the water softening section of your local water plant to learn of materials and methods used. Compare with materials used in home method of water softening.

5. List by name laundry aids you use in doing your own laundry, indicating reason for choice and method of using each.

6. Become familiar with methods of stain removal. Use samples of white cotton material stained with blood, salad oil, chocolate, ink, and lipstick, and apply desirable method for removal. Use bleach if necessary; justify use.

4 BASIC MECHANICS

THE NUMEROUS SMALL PIECES of kitchen equipment used to prepare food are frequently classified as *tools*. They can, moreover, be regarded as devices for the application of force.

FORCE. From a physics point of view such equipment may be reduced to two fundamental types: the lever and the inclined plane. The wedge, wheel and axle, and gear combinations are modifications or combinations of these two. In their application of the force principle, we find that usually a small effort is made to overcome a fairly great resistance. The ratio of resistance to applied effort, or $\frac{\text{output force}}{\text{input force}}$, equals the AMA (actual mechanical advantage) of the device.

THE LEVER. The lever is a rod or bar supported on a pivot, called the fulcrum, by means of which the resistance R is overcome by a force F. In practice the pivot is usually a line perpendicular to the lever. The distance of the fulcrum to F may be designated by distance F, and that from the fulcrum to the resistance as distance R; then $F \times$ distance $F = R \times$ distance R, or $\frac{R}{F} = \frac{dF}{dR}$. The ratio $\frac{dF}{dR}$ is called the TMA (theoretical mechanical advantage). AMA and TMA are equal only when any loss caused by friction in the device is negligible.

The fulcrum may have three possible arrangements. It may be between the force and the resistance, as it is, for example, in a pair of scissors. Or the resistance may be between the fulcrum and the force as it is in the nutcracker, the device used to pry the cap from a soft drink bottle, and the French cook knife. Or the force may be between the fulcrum and the resistance, as it is when the hand is used to lift and carry pots and pans from the sink or mixing center to the range. In the latter case the forearm acts as the force, the object in the hand as the resistance, and the elbow as the fulcrum. In summary the arrangements are: force–fulcrum–resistance; force–resistance–fulcrum; and resistance–force–fulcrum.

Eggbeater wheels are an application of the lever, one wheel turned by the force *F* of the hand, the pinion wheels turning the blades of the beaters which represent the resistance. The wheels are mounted on separate axles; the turning of the drive wheel causes the pinion wheels to turn. Each wheel behaves like a lever with the radius of each wheel representing the distance *dF* or *dR*. Sometimes two wheels are connected by gear teeth; this was represented by the intermeshing of the drive and pinion wheels of the beater. The ratio of the two sets of gear teeth to each other may increase the speed of action or reduce it. Interlocking gears are used to turn the wringer of the conventional washer and to record kilowatthour and gas consumption on their respective meters.

THE INCLINED PLANE. The inclined plane finds application in the wedge; for example, the tapering sides of knives. The wedge is a double inclined plane and knives are combinations of wedge and lever. Another example is the screw, so commonly used to fasten parts of tools together. The screw thread may be considered an inclined plane wrapped around a central cylinder or core.

THE PULLEY. A third type of device for the application of force is the pulley. A belt usually connects the force and the resistance. In conventional washers a belt may connect the motor to the washer mechanism. In motor driven brush-type and motor driven agitator-type electric cleaners, the brush or agitator roll is connected by a belt to the pulley. A Venetian blind is raised or lowered by means of a rope over a pulley.

FRICTION. Although frictional losses in most lever devices are small, in other types of devices friction may be large enough to require consideration. Friction acts to oppose motion. In household tools and appliances, friction is reduced by the use of bearings and the use of lubrication. Bearings are of three types: sleeve, roller, and ball. In sleeve bearings the two surfaces slide across each other and friction is greatest. In the conventional washer the shaft of the wringer rests in a sleeve bearing where lubrication is almost prohibited because of possible damage to the materials. In

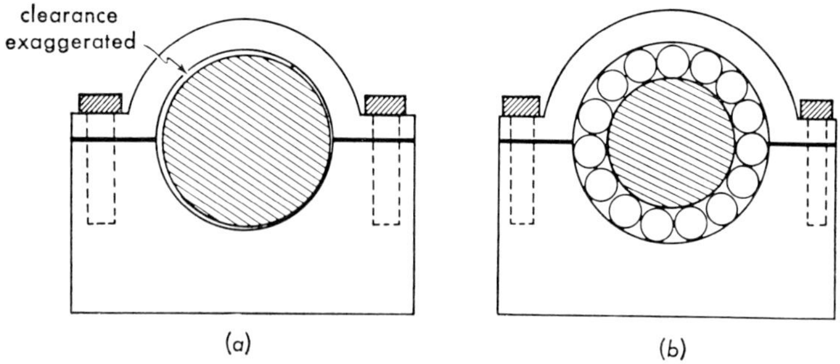

FIG. 4.1. (a) Sleeve bearing, showing shaft touching bearing over a considerable area. Sliding friction occurs. (b) Ball (or roller) bearing, showing shaft rolling on balls and balls rolling in the outer race. Only rolling friction is involved. *(From* Introductory Applied Physics, *2d ed. New York: McGraw-Hill, 1963. Used with permission of McGraw-Hill Book Company.)*

older wringers, the sleeve bearing was of wood and the homemaker often sprinkled soapy water over the ends of the wringer shafts to make them slick and thus able to turn more easily. At present, to minimize friction, the shaft and sleeve are usually made of two different metals. In roller bearings the contact between the two surfaces is along a line while in ball bearings only point contact is made; consequently, in both cases only rolling friction is involved. And this friction may be caused primarily by inelastic deformation of the rollers or balls and the race. A film of oil or lubricating grease further reduces the friction. The following table compares sliding and rolling friction.

TABLE 4.1. Coefficients of Sliding and Rolling Friction

Material	Coefficient of Friction
Steel on steel (dry)	0.20
Steel on steel (oiled)	0.06
Steel roller bearings on steel	0.003
Steel ball bearings on steel	0.0025

PROJECTS AND EXPERIMENTS

1. Diagram the three possible arrangements of force, fulcrum, and resistance. List at least three applications of each arrangement in kitchen tools.

2. Read in physics textbooks and write a discussion on each of the following:
    a. the use of lubrication in reducing friction
    b. advantages of roller and ball bearings over sleeve bearings
    c. a choice of alloys for sleeve bearings

5 HEAT: PRODUCTION AND TRANSMISSION

PRODUCTION. The production and transmission of heat follow certain physical laws. Heat is a form of energy, produced by the incessant motion of molecules, a type of energy known as kinetic energy. When heat is applied to a body, the velocity of the body molecules is increased and at the same time, therefore, the kinetic energy of the body molecules increases. The speed of the motion determines the temperature of the body; the greater the velocity, the higher the temperature.

Heat always normally flows from a hot body to a cooler one, or from a warmer area of the atmosphere to a colder one. When the direction of flow is reversed and the flow is from cold to hot, work is then performed. One body loses and the other body gains internal energy. The term heat energy is applied only to the transmission of energy, not to the internal energy of a body, not to a body in a state of equilibrium. When a body gives off heat, its internal energy is reduced; a body absorbing heat has its internal energy increased. When a part of a body is heated, the molecules are set into more rapid vibration; these molecules strike against adjacent molecules increasing their vibration, and the process continues until the whole body reaches the same temperature, i.e., temperature equilibrium.

Other forms of energy frequently act to product heat energy; for example, an electric current flowing through a wire against resistance will cause the wire to become hot. Coal or natural gas unites with the oxygen of the air to produce sufficient heat to cook food or warm a home. Without solar energy life on the earth would not be possible.

As explained in chapter 1, all matter is made of molecules and exists in one of three different states, solid, liquid, or gaseous. Usually solids have a definite volume and shape. The molecules of a solid are close together and are held in fairly constant relation to one another by cohesive forces. They tend to oscillate about fixed points. Heat can cause a solid to increase in length, volume, and area. A liquid has definite volume only, a volume which may expand under heat. The molecules in a liquid are almost as close together as they are in a solid, but have no fixed position and may slide over one another as they move around. A gas has neither definite volume nor shape; it always fills the container in which it is inclosed. If part of the gas escapes from the container, the remaining portion will immediately expand in volume to occupy all the space. For this reason it is rather difficult to obtain a perfect vacuum. In a gas the molecules are relatively far apart.

Heat can bring about a change in materials. For example, during the cooking process heat reduces the hardness of such vegetables as potatoes, carrots, and broccoli, and, during roasting, the toughness of meats. Heat increases the plasticity of fats, the malleability and ductility of metals. Heated gold and silver may be hammered into thin sheets, heated aluminum and copper may be drawn into wires.

TRANSMISSION. Heat may be transmitted from a place of higher temperature to a place of lower temperature by any one of three methods, conduction, convection, or radiation.

CONDUCTION. Conduction is the transfer of heat from one molecule to another in the same or different bodies with no visible motion of the heated bodies. The molecules, however, are in violent motion, passing on energy from one molecule to another. In general, metals are good conductors of heat, aluminum and copper the best; stainless steel is not so good, tending to have spotty areas rather than even conductivity. Ability of a material to conduct heat, called its thermal conductivity, is denoted by the letter *K*. The amount of heat conducted through an appliance depends not only on the thermal conductivity of the material from which it is made, but also on several other factors given in the following equation used to determine heat transfer:

$$H = \frac{K\,A\,T\,(t_2 - t_1)}{L}$$

where

$H =$ heat in Btu
$K =$ heat conductivity of the material
$A =$ area of material in square feet
$T =$ time in hours
$t_2 - t_1 =$ difference in temperature (F) between the two sides of the material
$L =$ thickness of material in inches

This formula for linear heat flow applies only if the cross-sectional area is uniform and does not vary from one place to another.

It is very evident that the size of the area through which the heat must pass and the degree of temperature difference between the two sides will affect the amount of heat transmitted, and the distance the heat must travel in passing from one side to the other will tend to slow the speed with which it moves. The ratio of temperature difference to distance is known as the heat gradient. In a combination refrigerator-freezer the temperature gradient of the food compartment may be 32 F to $2\frac{1}{2}$ inches (70 F room temperature to 38 F inside the food chamber, the refrigerator wall $2\frac{1}{2}$ inches thick), but if the freezer section is held at zero F, the temperature difference will be 70 F and, therefore, to keep the inward flow of heat less rapid the insulation must be increased to 3 or $3\frac{1}{2}$ inches.

A problem will illustrate the application of this equation: When used at sea level, the bottom of an aluminum saucepan, 8 inches in diameter and $\frac{1}{8}$ inch thick, placed on an electric unit, has a temperature of 214 F when the water is boiling. How many Btu per minute must the electric unit supply to the saucepan?

$$K \text{ for aluminum} = 1{,}480$$
$$A = \pi r^2 = 3.1415 \times 16 = 50.2656 \text{ inches}^2 = 0.349 \text{ ft}^2$$
$$T = \frac{1}{60} \text{ minutes}$$
$$t_2 - t_1 = 214 - 212 = 2$$
$$H \text{ (Btu)} = \frac{1{,}480 \times 0.349 \times 1 \times 2}{\frac{1}{8} \times 60} = \frac{1033.04}{7.5}$$
$$= 137.7 \text{ Btu/min.}$$

Wood and glass are only fair conductors; still air held in tiny pockets of a porous substance is a poor conductor. That is why Fiberglas and urethane foam, inclosing as they do many tiny dead air spaces, make good insulators for ranges and refrigerators. Cork was once used and has also a low *K* factor.

The heat conductivity of certain materials used in utensils, as insulation in appliances, or in the walls of a house are given in Table 5.1. When the *K* value is high, as in the case of metals, the material is a good conductor of heat; insulators have a low *K* value.

Aluminum and copper are especially good conductors. Stainless steel is not, but certain manufacturers of steel utensils have

TABLE 5.1. Heat Conductivity of Certain Materials
Btu/hr/sq ft/°F/inch

Material	K
Copper	2,640
Aluminum	1,480
Iron (Cast)	350
Insulating Fiber Board	0.35
Cork (Granulated)	0.28
Rock Wool	0.27
Kapok	0.24
Air (Still)	0.168

welded copper or aluminum on to the bottoms of the utensils, which has improved their conductivity.

CONVECTION. Convection is heat transfer by the motion of the medium as a whole. The medium is usually air or a liquid which moves from one area to another carrying the heat with it, frequently called a convection current. The winds act as natural convection currents, bringing icy air from northern Canada to the states during winter months. In contrast, the Gulf Stream carries heat from the tropics to northern shores. Heat is carried to the food in a gas oven by convection currents; in a refrigerator heat given off from food moves by convection to the freezing unit. As has been noted, heat always moves from a warmer to a colder area, when water is heated in a saucepan from room temperature to boiling, convection currents move through the liquid; the currents are formed by the expansion of the liquid by the heat, and therefore its density is decreased. The lighter liquid rises and the less-heated, heavier liquid flows downward to take its place.

The condenser of a refrigerator is frequently fastened to a finned or latticed panel at the back of the box, around which air circulates and carries the heat by convection away from the coils and into the room. Convection currents bring the heated air from a hot-air furnace through ducts to the various rooms in the house, while at the same time the cooler air returns to the furnace through other ducts.

RADIATION. Radiation, the third method of heat transfer, is an electromagnetic wave motion, similar to the passage of light through space. In fact, radiant heat travels with the same speed as light, 186,000 miles per second, but is of longer wavelength. Like light, radiant heat travels in straight lines and can be reflected and absorbed as well as transmitted. Shiny, light-colored surfaces tend to reflect light; rough, dark surfaces to absorb it. Good absorbers are also good radiators. Our sun is the most important source of radiant heat, supplying slightly more than 6 Btu per sq ft per minute to the earth during the hours when the sun's rays are perpendicular to the earth's surface. The following description illustrates very graphically the heat we receive from the sun, "In three minutes the earth receives energy equivalent to all the coal consumed in the United States of America during an entire year. And a single acre of the earth's surface, when full sunlight falls perpendicularly upon it, receives an amount of energy each hour which would cost nearly $200, if paid for at the usual electric rates!"[1]

In the equipment field electric broilers and rotisseries are sources of radiant heat. Gas ovens, as well as electric ovens, once the interior dark surfaces become heated, also give off radiations, probably responsible for much of the browning of food. The electric toaster browns the bread by heat radiation.

MEASUREMENT. Heat as it has application to measurements in household equipment is measured in Btu (British Thermal Unit). A British Thermal Unit is the amount of heat that will raise the temperature of one pound of water one degree Fahrenheit. Water is chosen as the unit because it has the highest heat capacity. All other substances have fractional values. In work in foods and in nutrition, temperatures are usually measured in calories, the amount of heat needed to raise or lower the temperature of one gram of water one degree Centigrade. However, instead of using the gram-calorie in food and nutrition calculations, the usual practice is to measure energy in the kilo-calorie (Calorie with a capital C), the energy given off by a kilogram of water in cooling one degree Centigrade. The two measurements, Btu and Calorie, differ only in the temperature and mass units used. Since one pound equals 454 grams and 1 degree Fahrenheit equals 1.8 C, one Btu equals 454 ÷ 1.8, or 252 calories.

A distinction must be made between heat and temperature. A thermometer measures the degree of intensity of heat a substance possesses. Without the use of a thermometer, scale temperature is at best a relative term; we speak of cold and hot according to the way our sense of touch reacts.

Two thermometer scales are in common use, the Fahrenheit

1. Lester T. Earls, *Brief Course in Physics* (Dubuque, Iowa: Wm. C. Brown, 1968), p. 89.

scale designed by a man of that name in 1714, and the Centigrade scale, developed by Celsius in 1742. Fahrenheit started to call the lower point of the scale 0 and the upper point, 100 degrees, using the temperature of an ice-salt mixture for 0 and body temperature for 100, but later methods used the freezing and boiling points of water for these two points at a standard pressure of 76 cm of mercury. When the Fahrenheit scale was compared, the freezing point was 32 degrees on the Fahrenheit scale and the boiling point 212 degrees. The Centigrade scale suggested by Celsius used 0 and 100 for freezing and boiling points. In 1948 the Ninth General Conference on Weights and Measures voted to use the name Celsius instead of Centigrade. This designation has been widely adopted in Europe, but not so generally in America (Fig. 5.1).

In comparing the two scales it is noted that 180 F (212 — 32) is equivalent to 100 C, or one degree Centigrade represents a change

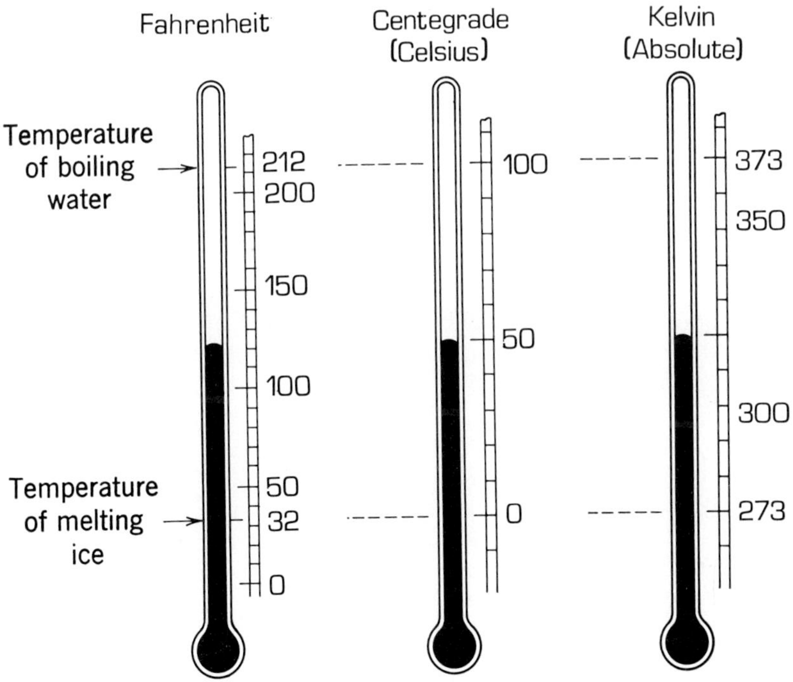

FIG. 5.1. Comparative scales of Fahrenheit, Celsius, and Absolute thermometers.

in temperature of 1.8 F. To state it in other terms, 100 : 180 = 5 : 9. Either method may be used:

1. To change Fahrenheit to Centigrade, subtract 32 from the given Fahrenheit temperature and divide by 1.8 or multiply by 5/9.
2. To change Centigrade to Fahrenheit, multiply the given Centigrade temperature by 1.8 or 9/5 and add 32.

Written as equations these methods are

$$C = \frac{F - 32}{1.8} \quad \text{or } 5/9\ (F - 32)$$

$$F = 1.8\ C + 32 \text{ or } 9/5\ C + 32$$

Examples:
Change 95 F to Centigrade

$$\frac{95 - 32}{1.8} \text{ or } 5/9\ (95 - 32) = 35\ C$$

Change 80 C to Fahrenheit

$$F = (1.8 \times 80) + 32 \text{ or } (9/5 \times 80) + 32 = 176\ F$$

When one desires to compare readings below zero, one must remember that the freezing point on the Fahrenheit scale is 32. For example, in determining what Centigrade degree will correspond to —31 F, the —31 is 63 degrees (32 + 31) below the freezing point and $\frac{63}{1.8} = 35$, i.e., 35 degrees below the freezing point, or —35 C. At absolute zero, 273 below the zero of the Centigrade scale and —459 F, all extractable heat has left the body.

In taking measurements in household equipment and also in engineering, the Fahrenheit thermometer is commonly used. The liquid-in-glass thermometer usually contains mercury or alcohol, mercury for the higher temperatures above the boiling point of alcohol, 171 F, and alcohol for temperatures below the freezing point of mercury, —38 F. Glass is a fairly poor conductor of heat, so the glass bulb of the thermometer is made as thin as possible without danger of breaking and the stem through which the liquid rises is evacuated and sealed to eliminate air pressure on the surface of the liquid as it rises and falls in recording temperatures.

A number of other types of thermometers and temperature devices have been developed to meet special needs, but they are not commonly used in household equipment. The thermocouple, however, is used and is very desirable for obtaining temperatures within appliances where thermometers cannot be used or if used, would require the opening of doors and so change the reading.

About 1820 T. J. Seebeck noticed that if two different metal wires were joined to form a closed loop, and then the two junctions were held at different temperatures, an electric voltage was produced which caused a current to flow through the wires. When used in equipment measurements the thermocouples of copper and iron or of iron and constantan (an alloy of 55% copper and 45% nickel) are connected to binding posts on a potentiometer. Wires of identical materials connected to the posts inside the instrument are joined to form a second junction. The junction inside the instrument remains at a constant room temperature. The temperature of the outside junction, which may be several feet away inside an oven or inside a refrigerator, can be read accurately on the potentiometer scale. The voltage produced has been calibrated in temperature units on the scale so that it may be read directly. The very small size of the junction facilitates its ease of use in many different experiments. Some potentiometers are portable and some are recording, giving a permanent record. More than one thermocouple may be used at the same time. Recording potentiometers frequently provide for the attachment of 12 or 16 thermocouples and the portable instrument may be used with a rotary switch, by means of which several thermocouples are placed in different areas and read consecutively.

CONTROL

INSULATION. To keep heat where it belongs, insulation is frequently used. Insulation is of several different types, depending upon the place where it is used. In heating appliances it is often Fiberglas. See Chapter 2 for full discussion of Fiberglas and other insulating materials.

Materials used on small electrical appliances and utensils to make them easier to handle are frequently plastic in nature. The most commonly used are the phenolics which form handles, feet, bases, and cradles on pots and pans, toasters, coffee makers, and frypans.

THERMOSTATS. Insulation inhibits heat flow. Thermostats, on the other hand, control the intensity of the heat, maintaining it at as nearly a constant temperature as possible. Thermostats are used in most small electric appliances, often referred to as electrical housewares in articles in women's magazines.

Appliances are distinguished from utensils. Appliances have the means of operation inherent in themselves, but utensils must be used with a source of power, usually on a range surface unit, on a burner, or in an oven, or be manipulated by the hand of an operator.

Thermostats control the heating of toasters, roasters, grills, rotisseries, coffee makers, electric frypans. They control also heat in range ovens, automatic range surface units and burners, and temperatures in refrigerators, furnaces, and air conditioning systems. Solid state components, discussed in chapter 9, are now used frequently as controls on many small appliances and on ranges, washers, and dryers.

Types. Thermostats used in household equipment are of three types: hydraulic, bimetallic strip, and probe tube.

HYDRAULIC. Used in ranges and refrigerators, this type consists of a small hollow tube, enlarged at one end for 6 to 10 inches. This enlarged end is installed in the oven or refrigerator with the other smaller end connected to the control switch. The tube contains a fluid that expands with heat, opening the electric circuit in the electric range and partly closing the gas valve in the gas range. As the ovens cool, the liquid contracts and reverses the action. In the refrigerator the opposite action takes place; the circuit is closed when the temperature rises, and the fluid expands and opens the circuit as the refrigerator cools.

BIMETALLIC STRIP. The bimetallic strip is used in small electric appliances. As the name implies, it is made of two strips of different metals, often iron and brass, fastened together at a certain temperature at which they lie flat. Since different materials expand at different rates when the temperature increases, one of the metals will expand more than the other and the strip will bend and break the electric contact. The temperature drops, the metals cool and assume their original shape, the contact is closed again. This increase in length per unit-length, per degree-rise in temperature is known as the coefficient of linear expansion.

PROBE TUBE. The probe tube is commonly used in electric housewares when one connecting plug is provided for a family of small appliances (Fig. 5.2). The tube contains silicone, a fluid that is sensitive to slight changes in temperature and is considered more accurate than the bimetallic strip. Appliances supplied with a probe tube usually can be submerged completely in water, a fact that greatly increases the ease of cleaning them.

SPECIFIC HEAT AND LATENT HEAT. Two other applications of heat are found in the operation of the refrigerator: specific heat and heat accompanying a change of state.

Specific heat is the quantity of heat measured in Btu needed to change one pound of a substance one degree Fahrenheit. Water is taken as the standard (it is also, you will remember, used in defining the Btu) because it has the highest heat capacity of any commonly used substance. The specific heats of all other materials are less than one. The higher the specific heat of a substance, the larger the percentage of water it contains. The specific heats of foods usually refrigerated are given in Table 5.2. It will be noted that foods of high water content have higher specific heats.

Heat accompanying a change of state, known as latent heat, cannot be measured by a thermometer, for there is no change in temperature. When ice changes to water, the temperature remains

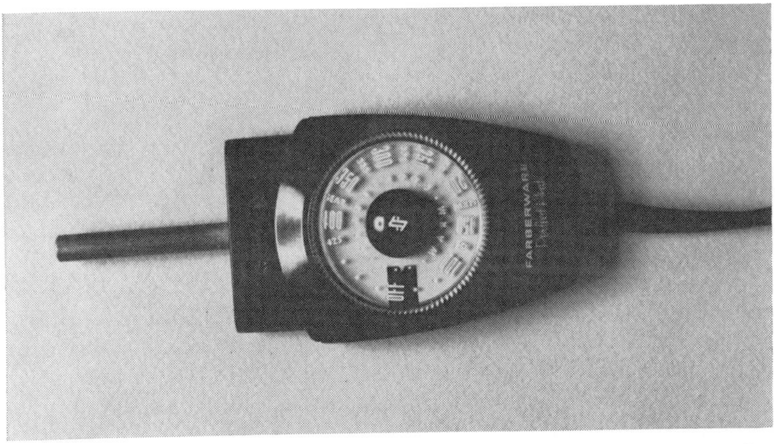

FIG. 5.2. The probe tube, the type of thermostat frequently used with a group of small appliances, in which the control is in the plug. It can be used interchangeably with several appliances and only one need be purchased, unless more than one appliance will be used at the same time.

TABLE 5.2. Specific Heats of Common Foods

Food	Sp Ht	Food	Sp Ht
Beans (Green)	0.92	Fish	0.82
Berries	0.89	Poultry	0.80
Oranges	0.89	Eggs	0.72
Peas (Green)	0.80	Ice Cream	0.90
Bacon	0.55	Milk	0.90
Beef	0.75	Watermelon	0.92

the same until all the ice is melted. Melting is known technically as fusion. One pound of ice in melting will absorb 144 Btu, i.e., the latent heat of fusion of ice is 144 Btu per pound. Expressed in Calories, the heat of fusion is 80 Calories per gram. The ratio of 80 to 144 is again 5 : 9. When water boils and changes to steam, one pound of water will require 970 Btu of heat to bring about the change, or the latent heat of vaporization of water is 970 Btu per pound.

TABLE 5.3. Specific Heats of Certain Metals and Glass

Metal	Sp Ht	Substance	Sp. Ht
Aluminum	0.22	Glass	0.21
Copper	0.093	Water	1.00
Zinc	0.093	Ice	0.51
Iron (Steel)	0.115	Steam	0.48
Tin	0.055		

In a similar manner, substances known as refrigerants used in refrigerator units to absorb the heat from foods, undergo changes of state, from a liquid to a gas in the evaporator (freezing unit) and from a gas to a liquid in the condenser. $Freon_{12}$ is the refrigerant widely used. It has a heat of vaporization of 69.5 Btu per lb. When a solid changes to a liquid or a liquid changes to a gas, heat is absorbed. When a gas changes to a liquid or a liquid changes to a solid, heat is given off. In the refrigerator the evaporator in which heat is absorbed is either in the food compartment itself, or behind the food compartment wall from where a fan circulates the cold air. The condenser, however, from which heat is given off, is outside the refrigerator box, either at the back of the chest or below it with provision made for the circulation of the warm air away from the chest. The fact that water in changing to ice gives off heat explains why areas located near fairly large lakes may have a moderate temperature in the winter until the body of water is entirely frozen. Islands surrounded by water often enjoy uniform temperatures, too.

In designing and constructing a refrigerator, the engineer must know what common foods will be stored in the food compartment, through what temperature range they will be cooled, and how much refrigerant must circulate to bring about the desired cooling.

Suppose, for instance, a homemaker buys at the grocery the following foods. By the time she reaches her home the foods have a temperature of 70 F. The temperature inside the food chamber is 38 F. How many Btu of heat must be removed in cooling the food to the 38 F?

TABLE 5.4. Btu Removed in Cooling Food

Food	Sp Ht	Btu (°F)
Berries, 2 lb	0.89	1.78
Bacon, 1 lb	0.55	0.55
Beef, 5 lb	0.75	3.75
Eggs, 2 doz (1 doz = 1.34 lb)	0.72	1.93
Milk, 2 qt	0.90	3.78
Watermelon, 15 lb	0.92	13.80
Total		25.59

The total Btu 25.95 $\times$ 32 (the difference in temperature, 70 — 38) equals the total Btu to be removed, 818.88. Thus, 818.88 $\div$ 69.5 (heat of vaporization of $Freon_{12}$) = 11.78 lb of $Freon_{12}$ needing to circulate. If 1.5 lbs circulate during each cycle, approximately 8 cycles would be required to bring about the cooling of this amount of food. If you were at your summer camp and using an ice refrigerator, you would divide by 144, the heat of fusion of ice and 5.7 lbs of ice would melt to bring about the cooling. In earlier years when ice chests were commonly used, the homemaker was often dismayed at the rapid melting of the ice and would wrap it in brown paper or newspaper or even in a special insulating blanket purchased at the store, not realizing that only as the ice melted would the food be cooled. However, it is true those old ice boxes were very poorly insulated and much heat came in through the walls.

The presence of impurities in water lowers its freezing point. A good mixture for freezing ice cream or sherbet in a hand-operated freezer is obtained by adding salt to the ice and so lowering the point of fusion. Impurities also raise the boiling point. A liquid boils when the pressure of the molecules escaping from the surface of a liquid equals the saturated vapor pressure within the liquid itself. This explains the difficulty of cooking foods on surface units at high altitudes where low pressure prevails. Under such conditions pressure saucepans are found to be of great value.

Usually, as has been seen, a solid changes to a liquid and it in turn to a gas. Some substances change directly from a solid to a vapor without going through the liquid state. This is known as sublimation. Examples are iodine crystals, moth balls, dry ice (solid carbon dioxide), frozen sheets on a line in a winter breeze.

To summarize the effects of heat: heat can cause chemical changes—reduction and oxidation; a change in electrical resistance, the generation of emf at the junction of two dissimilar metals of a thermocouple; a change in the incandescence of a wire, even resulting in the production of light; changes in the viscosity of a substance; changes in size, temperature, and state (toughness, plasticity, ductility, and malleability), and change in hardness of a body.

PROJECTS AND EXPERIMENTS

1. Convection currents. Bring a saucepan of cold water to the boil. Watch formation of steam bubbles and convection currents.

2. Radiation. Turn on electric heater. Note direction of heat flow. Note absorption of heat by article placed in front of heater. Place second article behind first, and note effect.

3. Conduction. Use rods of copper, aluminum, and stainless steel. Place one end of each rod in source of heat so that all receive heat equally. (Try placing rods between grids of waffle iron.) Let opposite ends rest on block of paraffin. Note at end of which rod paraffin is first melted.

4. Heat and cold are relative terms.
   a. Place one hand in cold water, the other in hot water for about one minute. Then place both hands in warm water. What is the sensation?
   b. Place hand on samples of carpet, silk, cotton, vinyl tile, silver spoon of same temperature (as a result of occupying same area for several hours). Describe apparent temperature difference.

5. Temperature equilibrium.
   a. Mix equal quantities of hot and cold water of known temperatures. Determine temperature of mixture.

b. Weigh $\frac{1}{2}$ cup of hot water and 1 cup cold water in two separate glass containers of such size that the hot water container may be placed within the cold water container. Note temperature when equilibrium is reached. What is loss or gain in temperature of each mass? Compare with original weight of each mass.

6. Fill a flask one-half full of water. Boil until all air is driven from flask. Stopper flask and turn upside down. What happens when steam above surface of water condenses? Explain.

7. Observe operation of recording potentiometer used with baking operation in oven.

8. Oil approximately same sized saucepans of aluminum, stainless steel, and copper-bottomed steel, and flour lightly. Heat each in turn over same unit or burner at same heat setting. Note length of time required for flour in each pan to become equally brown. Draw conclusions as to conductivity of materials. Compare results with experiment 3.

9. Test efficiency of insulation of one electric and one gas oven. Heat each oven at 400 F setting for one hour. Test for hot spots on top, sides, and door areas.

6 ELECTRICITY

It is difficult to believe that the first electric light was manufactured less than one hundred years ago, in 1879, and electric appliances much later, the first electric range in about 1910. Today, the modern homemaker would scarcely know how to keep house without her many electric servants.

GENERATION AND TRANSMISSION

GENERATORS. Electricity starts its journey to the home at the generating plant (Fig. 6.1). The heart of this plant is the generator. A magnetic, single-phase generator consists of two main parts, the field magnets and the armature. Lines of force run from the north pole of one magnet to the south pole of the other. The armature rotating between the two poles cuts these lines of force and generates an electromotive force (emf) or voltage. Either the field coils or the armature may be stationary and the other part rotate, but in either case the lines of force are cut and voltage generated, 115/230 volts or preferably 120/240 volts. The ends of the armature coil are joined to slip rings against which brushes press; by this means the rotating coil is connected to the external circuit.

Depending on the connections of the armature windings, the generator may be a single phase or polyphase generator, i.e., two-phase or three-phase. In the single-phase generator the armature windings are connected in series and the separate emfs add together.

The above discussion has been concerned with the single-phase generator. In the two-phase generator two armature windings are mounted 90 degrees apart, making the emfs also 90 degrees apart in phase. In this case the emf is never at zero point in the cycle (Fig. 6.2). In the three-phase generator there are three sets of windings and emfs, 120 degrees apart. The three coils may form three individual single phase circuits or may be interconnected in a Y or delta form (Fig. 6.3). To prevent a short circuit, the three windings must be connected to the outside circuits in such a way

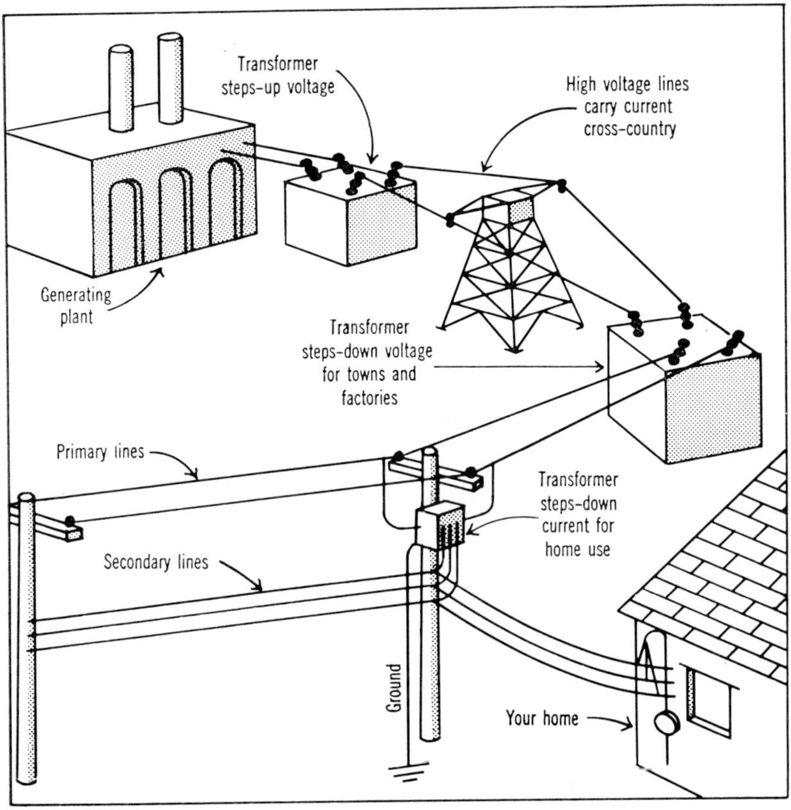

FIG. 6.1. Generation, transmission, and distribution of electricity. *(Westinghouse)*

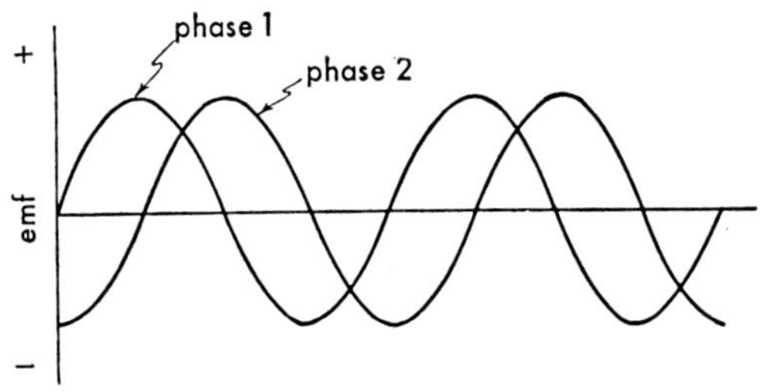

FIG. 6.2. Emfs of a two-phase generator are 90 degrees apart.

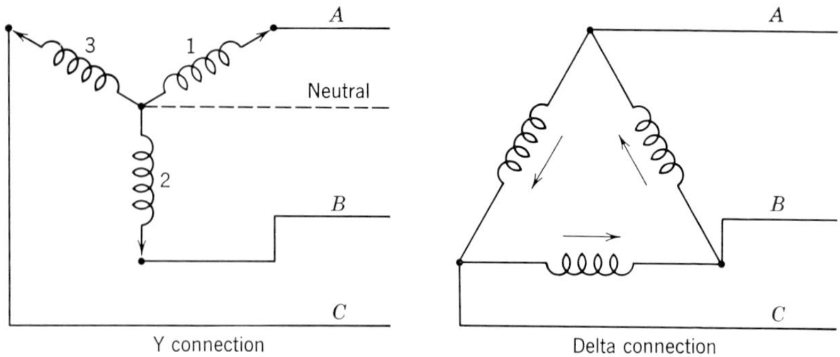

FIG. 6.3. Y and delta connections of a three-phase system. Current in all three windings must flow in the same direction.

that the current will flow in the same direction in each winding. In general, the three-phase generator is supposed to have superior operating characteristics.

CURRENT. All materials contain electrons, electricity-carrying particles. In such metals as copper and aluminum, the electrons, pushed by the emf, tend to pass from one atom to another, forming what is known as current. The generated emf forces the current through the conducting wires. The armature may be turned by water power if a stream has sufficient drop in level to turn turbines located within its depths to which the armature is connected. Rivers are frequently dammed to provide the needed drop in level and electric plants are built beside the dam. Niagara Falls, Hoover Dam, the Tennessee River series of dams, the Keokuk Dam in Iowa, and several on the Snake and Columbia rivers in the west are examples of hydroelectric power sources. If water is not available, steam from the burning of oil or coal may be used, but water is a cheaper source of power. Water and steam are sources of mechanical energy; therefore a generator is a means of transforming mechanical energy into electrical energy.

ALTERNATING CURRENT (AC). The emf and current thus obtained are alternating, changing from zero to a maximum in a positive direction, then diminishing again to zero; now cutting the lines of force in an opposite negative direction as they increase again to a maximum, and finally decrease to zero (Fig. 6.4). The

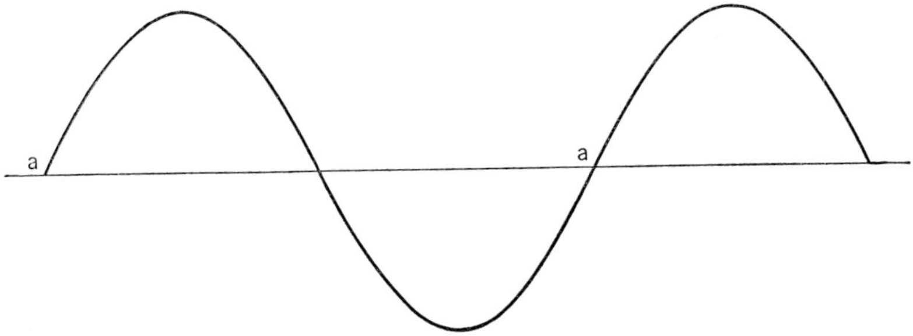

FIG. 6.4. Alternating current cycle.

curve, known as the sine curve, shows that the emf and current are continually changing in value. The distance from a to a on the curve is known as a cycle, and the number of cycles per second as frequency. The common alternating (ac) frequency in the United States is 60 cycles per second. In a few areas a 50-cycle frequency is used. Motor-driven appliances and those automatically controlled must be operated on the frequency for which they were manufactured.

In polyphase systems 120/208 volts are supplied instead of the 120/240 volts of the single phase system. In some states all state buildings, including all educational facilities, use 120/208 voltage. Those in charge of purchasing appliances for laboratories should know this fact to avoid making a wrong selection.

Builders of high-rise apartments often use this 120/208 voltage combination. If the builder installs electric ranges in the apartments, they will be the correct type, for he can order such ranges in volume directly from the factory.

Alternating current is generated in over 90 percent of the central power plants in the United States, and is, accordingly, widely used, chiefly because the voltage can be stepped up and down easily, and electrical energy can be transmitted over long distances at reasonable cost.

DIRECT CURRENT (DC). For certain manufacturing processes direct current is necessary, e.g., for electroplating and the charging of batteries. In such cases, if alternating current were used, the silver plated onto the cutlery or onto tea and coffee pots on the positive half of the cycle would be dissolved off during the negative half. Similarly a battery would be alternately slightly charged and discharged.

Direct current may be obtained from the alternating current generator by replacing the slip rings with a split ring called a commutator. The two segments of the split ring, insulated from each other, are attached to the two ends of the generator coil. The brush contacts change from one segment to the other as the current reverses direction in the rotating coil. This causes the current to flow in the same direction in the external circuit at all times.

CELLS AND BATTERIES. As has been noted, the electric generator employs the magnetic method of generating a difference in potential, the force that starts the electrons moving. This force may be generated by other methods also, by chemical, thermal, and frictional methods.

Volta, an Italian scientist, was the first to use the chemical method. He discovered that a difference in potential could be obtained by placing in a jar alternating metal plates of silver and zinc separated by paper sheets moistened by a solution of salt (NaCl). When the metal plates were connected outside the container, a current flowed from the negative zinc to the positive silver plate. This cell was used for about thirty years and was known as a voltaic cell. The voltage developed was small and therefore its use somewhat limited.

Then about 1830 Daniell constructed another cell which used a copper plate in place of the silver plate, in a jar filled with a dilute solution of sulfuric acid (H_2SO_4). This cell was used quite generally even as late as 1920. In operation the sulfuric acid breaks down into H_2 and SO_4 ions and the hydrogen tends to collect on the copper plate and cause polarization, which increases the internal resistance of the cell and consequently reduces the emf. To prevent polarization, an oxygen-producing depolarizer is added and the oxygen combines with the hydrogen to form water.

Cell is the name given to a single unit; when several cells are joined together, it is known as a battery.

PRIMARY CELLS

Dry. Cells are classified into primary and storage cells. The dry cell, although it is not really dry, is a primary cell. A zinc cylinder forms the wall of the cell and surrounds a carbon rod which is the positive electrode. Filling the space between the zinc and the carbon is a paste of ammonium chloride, manganese dioxide, small amounts of zinc chloride, and frequently some graphite to reduce the internal resistance of the cell, the paste mixture known as an electrolyte. The cell is sealed with wax to make it airtight and moistureproof. Two binding posts extend from the top. A new cell has an electromotive force (emf) of 1.5 volts. The emf diminishes with use, and gradually by just standing. The loss of emf may be retarded by placing the cell in the refrigerator or other cold place when not in use. It cannot be regenerated. Dry cells are commonly used in flashlights.

Mercury Dry. In the mercury dry cell the negative electrode is also zinc; the positive electrode mercuric oxide mixed with graphite. Potassium hydroxide and zinc oxide form the electrolyte. The mercury cell is frequently used in hearing aids because of its lower internal resistance and lower operating temperature. It does not deteriorate when not in use.

Standard. Another cell, the standard cell, is not used to supply energy, but in calibrating certain measuring and testing instruments. It is used in the potentiometer, for example, which is used for making precision measurements. The Weston standard cell will maintain a constant voltage for a long time. Before thermocouples are attached to the potentiometer for the purpose of measuring temperatures, the potentiometer is calibrated by connecting the standard cell into the circuit and adjusting the galvanometer scale reading until the pointer is at zero.

STORAGE CELLS. Storage cells use lead and lead compounds for the electrodes and dilute sulfuric acid for the electrolyte. When the cell is in use, lead sulfate ($PbSO_4$) gradually forms on both electrodes; at the same time water is formed:

$$Pb + PbO_2 + 2H_2SO_4 = 2PbSO_4 + 2H_2O$$

The cell can be regenerated by sending a current through it in a direction opposite to that of the original current. The emf is approximately 2.1 volts.

When a single cell does not supply enough current or voltage for a given need, several cells may be connected in series or in parallel to give increased emf or increased current. Six or eight cells connected together in series form the storage batteries used in automobiles.

BATTERY APPLIANCES. Cells or batteries are used to operate "cordless" equipment. At the present time a number of appliances that originally were operated by connection to an electric circuit are run from batteries and are often called cordless appliances. Among them are the electric knife, vacuum cleaner, toothbrush, electric shaver, flashlights, and certain radios, notably the transistor. Usually the cell is a nickel-cadmium type. The battery is rechargeable, the recharger an electronic device that changes the 120 volt ac current to low-voltage dc current needed for the batteries. The average time required for recharging is 12 to 14 hours. The battery is built into the piece of equipment permanently; can withstand severe vibration and shock; and under ordinary conditions will last as long as the appliance it serves.

OTHER GENERATION METHODS

THERMAL. When two wires of different metals are joined together at each end, and one junction is heated while the other remains cold, a difference in potential is developed at the two junctions that causes electrons to flow. This is a thermal method of generating emf. The two joined wires are known as a thermocouple. The voltage is too small for commercial application, but when the thermocouple is connected to a potentiometer, measurements of temperatures in ovens and refrigerators may be made.

FRICTIONAL. The frictional method of producing a difference in potential is obtained when two dissimilar materials are rubbed together; for example, hard rubber and woolen cloth. Sometimes when a person walks briskly across a wool rug and touches a metal object, tiny sparks may be seen leaving the finger tips, resulting in a tingling sensation in the fingers. This is a visible flow of electrons known as static electricity.

TRANSFORMERS. The use of electricity at a distance from the generating plant necessitates the use of a transformer to step up the voltage. A transformer consists of two coils of wire wound on a hollow iron core; the first coil which receives energy from the generator is known as the primary coil, the other coil to which the energy is passed is the secondary coil. The number of turns of wire on the secondary coil is a multiple of the turns on the primary coil and indicates how much the voltage has been stepped up. The ratio of the voltages on the two coils is equal to the ratio of the number of turns of wire on the two coils. Thus

$$\frac{\text{voltage on primary coil}}{\text{voltage on secondary coil}} = \frac{\text{number of turns of wire on primary coil}}{\text{number of turns of wire on secondary coil}}$$

For example, if there are 10 turns of wire on the primary coil and 200 on the secondary coil, and the voltage on the primary is 300 volts, the voltage on the secondary is 6,000 volts. Since the two coils are not in contact with each other, the emf in the secondary coil is produced by induction. The iron core aids by providing a path for a magnetic flux. By stepping up the voltage, the current is greatly decreased, the greater the step-up, the smaller the current, with the use of smaller wires to carry the current over the high lines across the country, and a corresponding decrease in energy loss.

As an example, when the voltage is increased 20 times, the current is reduced by the same factor or 1/20 of its former value, and the power loss in the line in the form of heat to 0.25 watts. Heat loss in the line is given by the equation

$$W = I^2 R$$

Assume that R, the resistance of the wire, a constant, is 100 ohms. Then

$$W = (1/20)^2 \times 100 = 0.25 \text{ watts}$$

At the edge of a distant city or town or at the entrance of a farm yard the process is reversed and the voltage is stepped down. The electrical system from the generating plant to the step-down transformer is known as the transmission system. Other step-down transformers are found frequently in other sections of the city or town to control the electricity used by a group of houses. You can recognize the transformer; it is usually a black rectangular box at the top of a service pole. The distribution system extends from the step-down transformer to individual houses.

CONDUCTORS. In order for current to flow, the wires must be good conductors. Copper, aluminum, and their alloys are the conductors commonly used, aluminum and copper alloys in the high lines, and copper in house wiring. Electrons are loosely held in the atoms of good conductors and pass easily from one atom to another.

INSULATORS. In contrast, electrons in insulating materials are held tightly and do not pass along from atom to atom. Insulators are used to keep the conductors in place on high line poles, to protect circuit wires in the home from contact with flammable material, and to prevent short circuits. Insulating materials are made of glass, porcelain, cotton, silk, rubber, plastics, and dry wood.

From the pole outside the home three insulated wires are brought to the house, occasionally underground in a waterproof duct, but more frequently overhead to knobs close under the eaves of the house. A recent tendency in new development areas is to put the wires underground. They are less apt to be damaged by tree branches and high winds, the cost of servicing is reduced, and the absence of poles greatly improves the appearance of the neighborhood.

SERVICE ENTRANCE. The three wires, known as the service entrance, must be, according to the regulations of the National Electric Code, at least 18 feet above any driveway and 10 feet above a sidewalk. The service entrance is rated in amperes and should be not less than 100 amperes; 150 and even 200 amperes are recommended. When an electric water heater is installed or the house is heated by electricity, the larger sizes are required. The three wires provide 120/240-volt service.

The three wire service conductors are dead-ended on strain insulators on the side of the house, and then brought down the side in a conduit pipe, a pipe with a very smooth, perhaps an enameled, inside finish, to protect the insulated wires from damage. At the present time the electric meter is commonly installed on the outside of the house to allow the meter man to read it each month without coming into the house; an added advantage from the homemaker's point of view, in bad weather or when she wishes to be away.

We have now traced the flow of electricity from the generating plant over the transmission and distribution systems to the service entrance of the house, down the conduit pipe and through the kilowatthour meter.

ELECTRICITY IN THE HOME

METERS. The dome meter has five dials, numbered from 0 through 9, and since they are interconnected, each dial is numbered in the opposite direction from the one before. The number the pointer has just passed is read and the numbers are recorded from left to right. The reading is in kilowatthours, the unit of energy. The reading from the previous month is subtracted from the present month's reading, and the kilowatthours used multiplied by the local rate to give the bill. Often companies have a sliding rate, so much for the first 50 or 60 kilowatthours (perhaps 5 cents per kwhr), a smaller rate for the next 100 kilowatthours (3 cents per kwhr), and all kilowatthours in excess of 150 for 2 cents per kwhr.

With the development of the rural electrification program, meters with the cyclometer type of register were used. This register is the same as that used on automobile speedometers. It can be read easily by the consumer and the kilowatthour reading mailed to the power company, thus reducing the company's operating costs.

It may be well to pause here and indicate the units of measure for the different electrical quantities, and the relationship between them.

UNITS OF MEASURE. Current is measured in amperes (symbol *I*), electromotive force (emf) or voltage in volts (symbol *E*), power in watts (symbol *P* or *W*), energy in kilowatthours (symbols kwhr or KWH), resistance in ohms (symbol *R*), power of motors in horsepower (symbol hp). Ampere, volt, watt, and ohm are all named after early scientists. The relationship between these quantities is expressed in the following true equations:

$$P \text{ or } W = EI,\ R = \frac{E}{I} \text{ or } E = RI \text{ or } I = \frac{E}{R}.$$

A kilowatt = 1,000 watts. Substituting the value of E in the equation $E = RI$ in the equation $P = EI$, we have $P = RI^2$. Horsepower is equal to 33,000 foot-pounds per minute, but in household equipment we usually use the relationship 1 hp = 746 watts. Many home appliances are of fractional horsepower: the electric beater, refrigerator, washer, dishwasher, and vacuum cleaner. Recently the tendency in the newer appliances is to rate them in wattage and more and more motor-driven appliances are so rated at present. Meters are available for measuring the various quantities.

AMMETER. The ammeter is an instrument for measuring current. The coil has very low resistance and most of the main circuit current does not pass through the coil, but through a shunt provided in the instrument. Sometimes more than one shunt is present, making possible readings on more than one scale. The ammeter is always connected directly into the line in series with the load. When current passes through the coil, the coil develops a torque that causes the needle to turn. The deflection of the needle over the scale indicates the exact ampere reading without any need for making corrections.

Because of its low resistance care must be taken not to allow too much current to flow through the instrument. When measuring large currents, a transformer is sometimes used and then the ampere reading must be multiplied by the transformer factor. The ammeter can also be injured by careless jarring (Fig. 6.5).

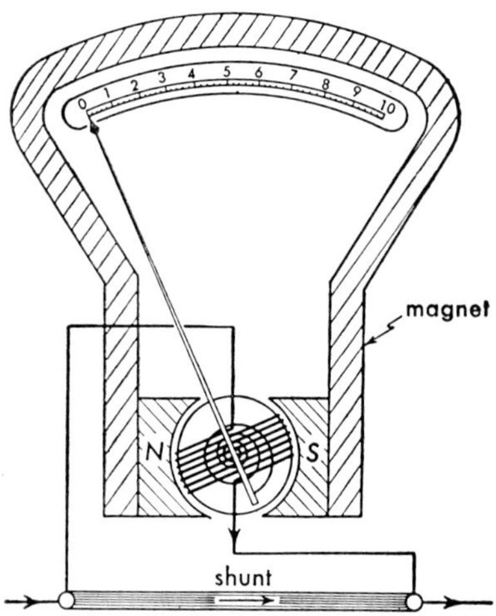

FIG. 6.5. Most of the current passes through the shunt in an ammeter. Only a small part passes through the moving parts of the instrument. (*From* Introductory Applied Physics, *2d ed., New York: McGraw-Hill, 1963. Used with permission of McGraw-Hill Book Company.)*

VOLTMETER. The voltmeter, connected in parallel with the load, is similar in appearance to the ammeter with a scale over which a needle passes to indicate the voltage drop taking place in the portion of the circuit indicated. The voltmeter is constructed with high internal resistance so as to divert as small a current as possible from the circuit into which it is connected. Voltmeters frequently provide more than one range of readings by supplying additional coils of varying resistance in series with the main coil. Connections should be made to the correct terminals depending on the voltage to be measured (Fig. 6.6).

Although alternating current changes in intensity value from maximum plus to maximum minus, the effective value of the current is equal to 70.7 percent of its maximum value. Similarly the effective emf is also equal to 70.7 percent of its maximum value.

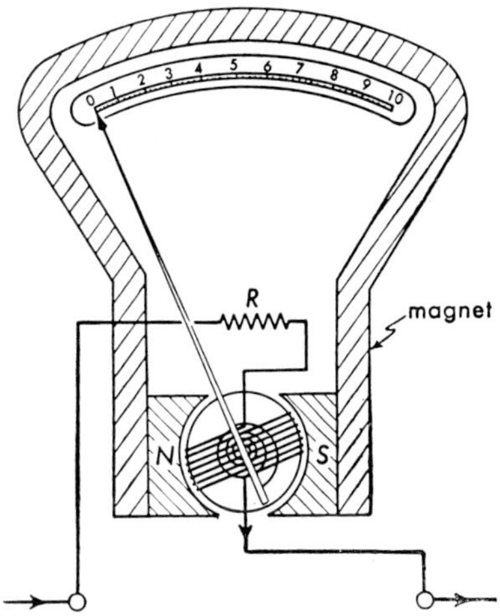

FIG. 6.6. Circuit of voltmeter. *(From* Introductory Applied Physics, *2d ed., New York: McGraw-Hill, 1963. Used with permission of McGraw-Hill Book Company.)*

Since the effective values are the usual measurements desired, the ammeter and voltmeter are calibrated to read effective values.

WATTMETER. A watt is the unit of power. The deflection of the coil, and therefore the needle, is proportional to the current times the potential difference. The wattmeter indicates the rate at which power is used at a given moment. Since the wattmeter combines the properties of the ammeter and voltmeter, it has two sets of binding posts for connection to the circuit. The winding representing the ammeter is connected in series with the load, the voltage coil in parallel with the load.

WATTHOUR OR KILOWATTHOUR METER. The kilowatthour meter used in laboratory experiments to measure energy is like the wattmeter in having two sets of binding posts, one for the current connection, the other for the voltage connection. The meter coil rotates continuously for the time the meter is connected to a load, the rate of rotation depending on the power consumed by the load. The cost of operation is based on the number of revolutions made and is recorded on a series of dials on the face of the instrument, like those found on the kilowatthour meter installed in the home. If the reading is too small for the scale to record, a student can count the rotations of the metal disc in the center of the meter, and then figure the consumption from the factor of the instrument. The consumption will be in watthours, which can be changed to kilowatthours by dividing by 1,000.

FUSES. The wires are brought from the conduit pipe through the wall into the basement or back hall or other desired area to the switch and fuse control box. Usually the first part of the system inside the house is the main switch, which may be opened if work needs to be done on any wiring within the building, or sometimes for safety while changing a fuse. The main fuses or circuit breakers come after the main switch. They are safety devices to prevent an overflow of current which always generates too much heat and could cause a fire. A fuse contains a thin bridge of low melting alloy that "blows," i.e., melts, when excess current flows through it. If not on the outside of the house, the meter is installed next. Finally, we have the power center with one fuse or circuit breaker for each 120-volt circuit and two fuses or circuit breakers to protect each 240-volt circuit.

When motors are first turned on there is an instant surge of current, before the flow becomes steady at a much lower rate. If several motor-driven appliances in the home were on the same circuit and started simultaneously a fuse would probably blow. To prevent this from happening a special fuse called a Fusetron has been developed, to take the first oversurge of electricity without blowing. Only when the excess current continues to flow will it blow like any other fuse. Apartment house owners use a third type, a Fustat, to prevent renters from replacing the correct size fuse with one too large. In this case an adapter for a given size fuse is screwed into the receptacle in the fuse box and cannot be removed. Only the proper size fuse can be used with it (Fig. 6.7).

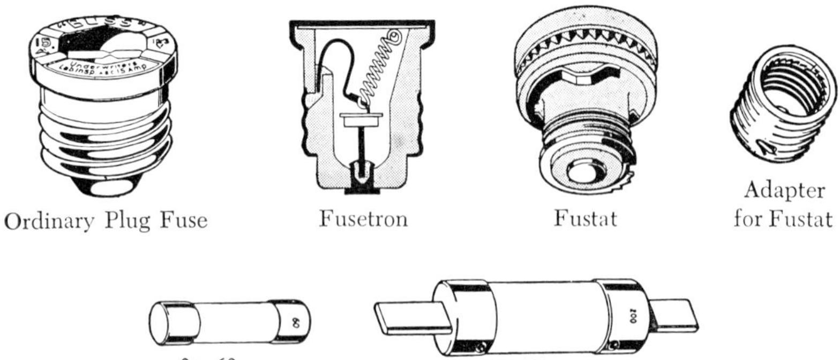

FIG. 6.7. Types of fuses, plug and cartridge. *(Bussman)*

CIRCUIT BREAKERS. In many modern houses circuit breakers are used in place of fuses. They are made of strips of two different metals fastened together, each with a different coefficient of expansion. When too much current flows through a breaker, one metal expands more than the other and causes the connection to be broken. When the trouble is righted, one need only flip a switch on the circuit breaker, and it is again functioning–much simpler certainly, than changing a fuse.

Two types of wiring are available, series and parallel. Reference already has been made to the series wiring of the armature windings in the single-phase generator. In series wiring the current is the same throughout the circuit; the total resistance is the sum of the separate resistance of each coil

$$R = r_1 + r_2 + r_3$$

The total voltage is the sum of the voltage drop across each coil, and each voltage drop is proportional to the resistance of that coil. In parallel circuits resistance is expressed

$$\frac{1}{R} = \frac{1}{r} + \frac{1}{r} + \frac{1}{r}$$

or the reciprocal of the resistance of each coil. The total current is the sum of the currents flowing through the different branches of the circuit and each branch current is inversely proportional to the resistance of the branch through which it flows. The voltage is the same throughout the circuit. These types of wiring will be better understood as we discuss the wiring in the home and the wiring of certain appliances used on the circuits.

WIRING. Three different types of circuits are found in the home: the general purpose circuit, the appliance circuit, and the individual circuit, each with its own size wire and each size wire protected by the size fuse designated by the National Electric Code.

The National Electric Code is a set of standard regulations drawn up by a committee of the National Board of Fire Underwriters (Fire Insurance Companies) and approved by the American Standards Association. They are minimal requirements for safety protection against fire or danger to life and are not necessarily adequate. Licensed electricians follow these regulations in wiring a home. If a fire occurs and it could be proved that an unlicensed person did the wiring, the owner might have difficulty in collecting the fire insurance.

The code specifies that No. 14 wire is the smallest wire size that may be used in a general purpose circuit and that No. 14 wire must be fused with a 15 ampere fuse. Fuses are rated in amperes. Since many electric companies try to supply 120 volts to their customers, and watts equals volts times amperes ($W = EI$), such a circuit would supply 1,800 watts (120×15). It is not possible to use the entire 1,800 watts without blowing a fuse, in fact, only approximately 80 percent should be used. That would be about 1,500 watts.

GENERAL PURPOSE CIRCUIT. The general purpose circuit has outlets for ceiling fixtures. Also, in living room, dining room, halls, and bedrooms are convenience outlets in the walls for the attachment of floor and table lamps, the electric cleaner, an electric hot pad or blanket, and even for occasional use of other small electric appliances. In the early years of home use, household appliances were of low wattage; an electric hand iron of perhaps 600 watts, a toaster, 550 watts, a percolator, 380 watts. Moreover, these three appliances were probably the only three electric appliances available in the home, and the possibility of overloading a circuit was very remote. In contrast, at the present time, two or three dozen electrical appliances are available and most families use a dozen or more regularly. And with the automatic controls to prevent excessive temperature, the modern appliance has a much higher wattage rating; a toaster now has a rating of 1,000 or 1,100 watts, the percolator, 800 watts, an iron, 1,150 watts, the frypan, 1,350 watts. Usually only one of these appliances can be plugged into an outlet on the general purpose circuit at any time. When an appliance is plugged in and the lights dim, it shows plainly that the outlet is on the same circuit as your lights. For this reason, many city electric codes supplement the National Electric Code and require that all new electrical installations in homes use No. 12 wire for the general purpose circuit. This size wire is fused with a 20 ampere fuse and provides 2,400 watts. Note that the size of wire varies inversely with the number, i.e., No. 12 wire is larger than No. 14.

It is considered desirable also to install more than one circuit in the walls of a room, one, perhaps, on two sides, and another on the other two sides with some outlets on each. Such an arrangement not only balances the load, but also prevents a blackout if a fuse should blow. This does not mean two separate circuits for each room. Each circuit may run through several adjoining rooms.

APPLIANCE CIRCUITS. Appliance circuits frequently are provided in a kitchen or breakfast alcove; perhaps also in one end of the dining room and in the family room, as well as in the laundry area. Such circuits use No. 12 wire, fused with a 20 ampere fuse, or preferably No. 10 wire, fused with a 25 ampere fuse, giving a wattage of 3,000. At least two appliance circuits should be installed in the kitchen area and an additional circuit in the laundry. Another method is to use a split-wired circuit. This is a three-wire circuit (Fig. 6.8) with 120 volts between either outside wire and

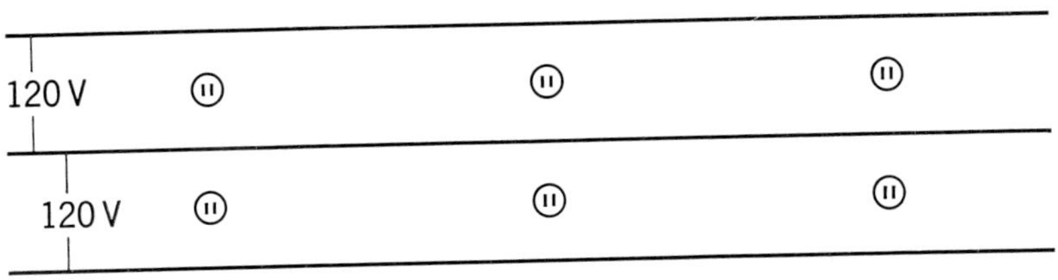

FIG. 6.8. Split-wire circuit, equivalent to two appliance circuits. Cost of one copper wire eliminated. No connection between the two outside wires.

the center wire, each side fused with a 20 ampere fuse, and therefore supplying two 2,400 watt circuits. Some outlets will be on the circuit on one side, others on the other side, allowing several appliances to be connected at one time without danger of blowing a fuse. This type of circuit is equivalent to two separate circuits but saves the cost of one copper wire.

INDIVIDUAL CIRCUITS. Individual circuits are of two kinds, those of 240 volts and those of 120 volts. The code requires any appliance using more than 1,650 watts to have a separate circuit. The electric range (14,800 watts or more), electric water heater (4,000 watts), and most electric clothes dryers (4,600 watts), all require 240 volt circuits and must have individual circuits, i.e., each must have its own circuit. Certain other appliances, the automatic clothes washer, dishwasher, furnace fan, and freezer do not have such high wattage ratings, but they are all motor driven. It was noted that a large charge of current is used when a motor first starts to operate. If two or more motor-driven appliances were on the same circuit, the first overflow of current could cause a fuse to blow. So it is recommended and some municipalities require that these appliances be on individual circuits. After this first overflow, the current is fairly small in amount. In general, motor-driven appliances use much less electricity than appliances that use electricity for producing heat.

All outlets on the general purpose and appliance circuits are wired in parallel. Then if a bulb in a lamp burns out or a toaster fails to operate, no other outlets on the circuit are affected. Only when a fuse blows is the entire circuit dead. Inexpensive strings of Christmas tree lights are occasionally wired in series. If you have had such a string, you know that if one bulb burns out all the other bulbs go out, too, and it is often quite a task to locate the faulty one. Now, most Christmas tree lights are wired in parallel, so that the bulb not operating can easily be replaced.

If you look at a range unit, you will notice that it usually has two tubes wound together in one of two forms; one tube may be in the center with the second tube on the outer edge, or the two tubes may each start at the center of the unit and occupy parallel circular positions, each ending at the outer edge. Each tube has a coiled insulated wire inside, the source of heat. A few units may be in the form of a single broad tube; in this case two coils of wire insulated from each other occupy the tube. Both types of units usually supply five heats, depending on how the two coils are connected to each other by the switch. For the highest heat the two coils are connected in parallel, for lower heats a single coil is used or the two coils are connected in series. If the two coils are of equal resistance, the connections may be as shown in Figure 6.9. Applying the two equations, $I = \frac{E}{R}$ and $W = EI$, and assuming that each coil has a resistance of 80 ohms, the various heats could be obtained as indicated below the wiring diagram.

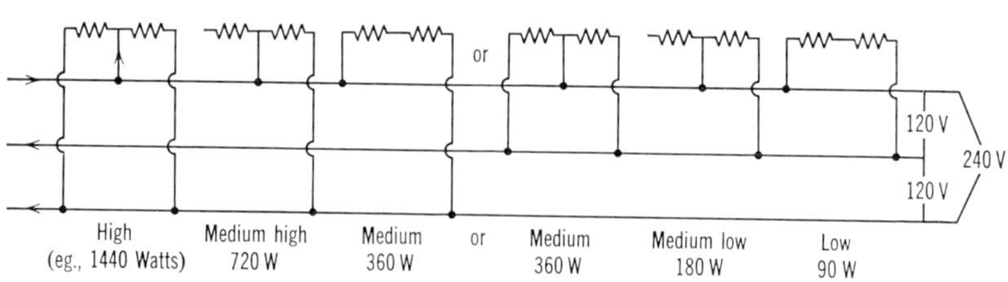

FIG. 6.9. Wiring diagram of a five-heat unit on an electric range.

When two equal coils are wired in parallel the resultant resistance is one-half the resistance of one coil.

When two coils are wired in series, the resultant resistance is the sum of the two resistances.

When only one coil is used, the resistance is the resistance of that coil.

The range manufacturer usually follows these general rules:

1. To obtain the highest heat the two coils are connected in parallel across 240 volts.
2. To obtain medium heat the two coils are connected in series across 240 volts or in parallel across 120 volts. The results are identical.
3. When the resistances in the two coils are equal, each wattage from "high" to "simmer" is half the preceding one.
4. To obtain more than five heats, the resistances in the two coils must be different; then the coils can be used separately to obtain additional heats.

Infinite heat units are also available, with a warm setting for the lowest heat and a high setting for the highest and an indefinite number of settings between. The early control device for these heats was a voltage control, but the latest control is current sensitive and regulates the wattage of the unit. A given control cannot be used with all units, but must be matched to a specific wattage.

The control is a double-pole switch that operates on the principle of a bimetal device. It is set manually by an internal cam which allows the unit predetermined on-off times by cycling to make or break contact with the line. Attached to the bimetal element is a heater connected in series with the surface unit; the heater cycles in unison with the unit. A pilot light is on continuously whenever the switch is turned to any "on" position.

At the high setting the unit is energized continuously; at the warm setting the unit is on approximately 5½ percent of the time, at low heat, 27 percent, and at medium heat, 35 percent. When the temperature of a setting is reached (except at the high setting), the bimetal control flexes, breaks contact with the circuit, and the unit cools. Below a certain temperature the bimetal again makes contact and closes the circuit. Cycling thus off and on the unit maintains the temperature desired. Heat is distributed evenly over the entire surface of the unit.

The unit may have a single coil, or be a multiple coil unit. Multiple coils are connected in parallel so that they function as a single coil. In using the infinite heat unit it is customary to start with the control on high until the food reaches the cooking temperature, then reset to a lower temperature for the remainder of the cooking time.

SWITCHES. Switches and convenience outlets complete the home wiring system. Switches are usually placed in the wall of the room on the latch side of the entrance door, 3½ or 4 feet above the floor. Unless entrances are very close together, switches should be placed at each door, so that it is possible to light a room as you enter and turn off the lights as you leave, without retracing steps in darkness. If a room has only one entrance, a single switch may be used, but if a light must be turned on or off at two entrances, three-way switches should be installed at each door. Two similar switches are placed at the bottom and top of stairs, one for each light, so that you may light the stairs ahead of you, eliminating your shadow, whether you go up or come down, and be able to turn off the unneeded light when you reach the desired level. Occasionally three switch positions are necessary. In this case two three-way switches and one four-way switch are used. If additional switches are required, a second or third four-way switch is placed between the three-way switches. Outdoor lights are controlled from the inside of the house by a switch beside the entrance door. Two or three switches may be installed under a single plate; concentrating switches whenever possible is desirable.

A switch is commonly the toggle type, easily manipulated by hand, recognized by the clicking sound accompanying its use. Mercury switches are increasing in popularity because they are noiseless and shockproof. They contain liquid mercury sealed inside the switch.

Sometimes a master switch is installed beside the bed in the master bedroom. Such a switch can be used to turn on or off all the lights in the house and floodlights in the surrounding grounds, a protection against prowlers. This switch may also make possible the turning on of a space heater in bedroom or bathroom in early morning before the house is warmed from the furnace.

CONVENIENCE OUTLETS. Convenience outlets are placed in the walls of all rooms and halls for the connecting of floor and table lamps, small appliances, and the vacuum cleaner. The receptacle outlet for the cleaner may be combined with a switch location to save stooping in making the connection. In living, dining, and bedroom areas the outlets are placed 15 to 18 inches above the floor, 12 feet apart on all walls where they need to be available. Most lamp and appliance cords are approximately 6 feet long, hence the spacing measurement. However an outlet should not be located in the center of a long wall where a large piece of furniture, a piano, a sofa, or a bed may be placed and interfere with the use of the outlet. In hallways a 15-foot spacing of outlets is adequate. In the kitchen, in addition to outlets for the range, if electric, and the refrigerator, outlets should be installed above the work counter at 4-foot intervals. It is common practice to install duplex outlets instead of single ones. Some local codes specify a ground connection on all outlets. The outlets for the dishwasher and clothes washer should always be grounded, and small appliances used near the sink should also be connected to a grounded outlet. However appliances with an open coil such as the toaster and space heater should not be grounded.

MOTORS. We have learned how generators change mechanical energy into electrical energy. Motors work in the opposite way, they change electrical energy into mechanical energy.

UNIVERSAL. Two types of motors are commonly used in household appliances, the universal and the induction motor. Each has a stationary part, the stator, and a rotating part, the rotor. In the universal motor the stator and rotor are connected in series, and current from the stator is fed into the rotor through brushes impinging on commutator bars. This motor may be operated on either ac or dc current, because the current in the armature and in the field reverse at the same time; hence the name universal. The advantage of the universal motor is its application when high speeds are required, and its ability to adapt automatically to the size of the load: high speed when the load is light; slower speed with a heavy load. It is used in hair dryers, electric shavers, food mixers, electric grinders, fans, vacuum cleaners, and in electrically operated sewing machines.

INDUCTION. In contrast, the induction motor has no contact between the stator and rotor. The stator section is connected to an ac supply and the current from it is induced into the rotating part. The magnetic field of the rotor reacts with the magnetic field of the stator to produce motion. It must be used on the frequency for which it was designed. It may be built for use on a three-phase circuit. Induction motors are found in washers, dryers, dishwashers, ironers, oil- and gas-operated heating furnaces, and air conditioners.

In certain appliances (for example, the automatic washer), the induction motor may have a capacitor attachment which uses less starting current than the conventional induction motor. The capacitor stores up an electric charge which supplies a high starting force known as *torque* to the appliance.

SHADED POLE. Another motor, the shaded-pole motor, is sometimes used in simple appliances such as hair clippers, can openers, knife sharpeners, and electric clocks. It is a very small motor, often 1/20 horsepower or less. It is reliable in operation and inexpensive, but is low in efficiency and tends to be noisy. The motor is made of laminated strips with a part of each pole split to hold a short-circuited copper strap called a shading coil.

The mechanical energy of the motor is transferred to the place of use by means of gears. For example, the agitator of a washer which rotates a third or half turn in one direction, then reverses and rotates the same number of degrees in the opposite direction is controlled by a rack and pinion gear. The wringer on a conventional washer is operated by a spur gear, which causes the wringer to turn continuously in one direction, or with a change of control, continuously in the other direction. Friction is reduced as far as possible by the use of bearings—roll, ball, or sleeve bearings (Fig. 4.1).

If a motor starts to throw off sparks, the appliance should be disconnected and the cause investigated. The brushes may be worn and if so they should be replaced. The commutator bars may show a buildup of a gummy accumulation of dirt mixed with oil; they should be cleaned. Motors should be lubricated if oil holes are present; use the correct type of oil and the correct amount, as indicated in the manufacturer's directions. Too much oil can be as detrimental as too little.

Another factor encountered in motors is counter or back emf. The magnetic field around the armature of a motor reacts with the lines of force between the field coils to cause rotation. As the armature rotates, it also cuts the lines of force set up in its own field (there is always a magnetic field around a conductor in which current is flowing), inducing an emf which is opposite in direction to the flow of current in the conductors. According to Lenz's Law the induced current has such a direction that the magnetic field set up by the current will tend to oppose the motion that produced it. In other words, the counter or back emf developed opposes any change in current flow and so maintains it at a fairly constant rate. If the speed of the rotor decreases, fewer lines of force in the magnetic field will be cut and the current will increase. As the current increases the magnetic field is enlarged, more lines are cut, more back emf develops, and the current decreases again. The back emf, which is 180 degrees out of phase with the applied voltage, therefore acts as a current control, increasing or decreasing as the load on the motor requires and holding the current within safe limits. We have already noted that too great a flow of current causes too much heat. Were it not for the control of emf, the current might easily build up beyond safe limits and burn out the motor.

The effect of drop in voltage must also be considered. A drop in voltage causes a heat appliance such as a toaster or electric frypan to heat more slowly and perhaps not reach optimum heat. Rural homes near the end of a line often have difficulties with drop in voltage when the other families on the line are using many appliances.

Voltage drop can be much more serious when it affects motors. The motor may try to start or struggle to reach operating speed without success. In consequence, excessive heat develops within the motor windings and the motor may burn out, unless it is equipped with a thermal overload protective device that opens the circuit.

POWER FACTOR. In ac circuits used in home wiring $W = EI$ because the voltage and current are in phase, i.e., the voltage and current rise to maximum values at the same time, fall to zero together, increase to maximum value in the opposite direction at the same time and together return to zero (Fig. 6.2). In the induction motor the voltage and current are out of phase, the current either ahead of or behind the voltage. In this case the true power in watts does not equal EI, which is known as apparent power. True power in watts divided by apparent power, EI, gives a decimal fraction, varying from 0.2 to 0.7, called the Power Factor, its value depending on the type of motor-driven appliance. Volts $\times$ amperes $\times$ the power factor gives the true value for watts. The power factor compensates for the difference in phase between the current and voltage. Because the power factor for a given appliance is not always known, motor appliances are preferably rated in watts.

Induction motors use three-phase ac very successfully. As commonly used in the home, electricity flows through two conductors to form a closed circuit. The three-phase ac requires only one conductor for each phase, because at any given instant the current in one phase is equal to the sum of the currents in the other two phases, but opposite in algebraic sign. Therefore using three-phase power obviates the necessity of obtaining current from the armature by means of a commutator and brushes.

HOME APPLIANCES

SOLENOID. Some appliances use the solenoid valve as part of their regulating mechanism. It is a coil of wire wound in a spiral around a soft iron bar. When an electric current is allowed to flow through the wire, the iron bar is drawn into the coil and becomes a temporary magnet, the coil and iron core together forming an electromagnet. The effect continues only as long as current flows through the coil. When current ceases to flow the bar is released and slides outside the coil. Therefore the primary advantage of the solenoid is its control of intermittent action.

The solenoid may be used to turn the water on and off automatically in filling the washer tub, in starting and stopping the agitator action, and the spin cycles. Similar intermittent action is used in factories where washers, ranges, and refrigerators are manufactured; to transfer scrap iron and steel to furnaces from stock piles in the plant yard; or to move large structural parts from one area to another.

NAME PLATE. Every appliance should carry a name plate, an indication that the manufacturer is proud of his product and stands behind it. The name plate is stamped into the metal of the appliance, usually on the bottom of it or on a separate attached plate. Occasionally it may be on a sticker, but that may come off and is rarely used, except for the Underwriters Laboratory (UL) seal. The name plate contains the name and address of the manufacturer, the model number of the appliance, the type of current (ac or dc), number of cycles, wattage rating, voltage on which the appliance should be used, and occasionally the amperes. When amperes are given, wattage may be omitted. On motor-driven equipment horsepower rating is sometimes found. Often also special information for the operation of the appliance is included, e.g., "attach to wall outlet only," or "do not immerse in water." Always look for the Underwriters Laboratories seal which indicates that the appliance has been tested for safety, safety against fire hazard and against current leakage that might cause a shock or even result in electrocution. All electric appliances should carry the UL seal. It is found on electrical equipment, on cords, on fuses, and wiring, the color varying with the type and size wire. On appliances it is usually the letters UL surrounded by a circle.

COST OF OPERATING. With the wattage rating of the appliance and a knowledge of the length of time it is used, the homemaker can determine the cost of operation; for example:

What will it cost to use an electric frypan rated at 1,350 watts for 20 minutes on Monday, 40 minutes on Tuesday, 30 minutes on Wednesday, and 30 minutes on Friday, at \$0.04 per kilowatthour, if the current flows 2/3 of the time?

$1{,}350 \text{ watts} = 1.35 \text{ kw}$

$20 \text{ min} + 40 \text{ min} + 30 \text{ min} + 30 \text{ min} = 120 \text{ min} = 2 \text{ hr}$

$1.35 \times 2 = 2.70 \text{ kwhr}$

$2.7 \times \$0.04 = \$0.108$

$2/3$ of $\$0.108 = \$0.072 = 7$ cents, cost of operation

A toaster's nameplate indicates 120 volts and 10 amperes. What will it cost to operate it 20 minutes a day during September at \$0.05 per kilowatthour?

Since $W = EI$, $W = 120 \times 10 = 1{,}200$ watts $= 1.2$ kw

20 minutes per day during September $= 30 \times 20 = 600$ min $= 10$ hr

$1.2 \times 10 = 12$ kwhr

$12 \times \$0.05 = \$0.60$, cost of operation

Problems are easily solved if fractions are expressed as decimal fractions.

A ½ hp washer (1 hp $= 746$ watts) was used 1½ hours a week. What would it cost to use it for a month at \$0.05 per kwhr?

½ hp $= 373$ watts $= .373$ kw

1½ hr $\times$ 4 (weeks in a month) $= 6$ hr

$.373 \times 6 = 2.238 = 2.24$ kwhr

$2.24 \times \$0.05 = \$0.112 = \$0.11$, or \$0.22, if we remember that the fractional horsepower motor is only about 50 percent efficient. Notice that motor-driven appliances in the same length of time use less electricity than heat appliances do.

SAFETY

Grounding. The electric current always seeks to reach the ground. Certain areas of our environment are in contact with the ground, the basement, garage floor, and patio; also through the cold water pipes, the kitchen sink and bathroom fixtures, the clothes washer and dishwasher. If by chance you are in contact with a defective appliance that has current leakage at the same time you are in contact with the ground, the current may flow through you, causing a severe shock and perhaps a fatal accident. Therefore all appliances used in these areas should be grounded. To ground an appliance use a three-prong grounded type plug attached to a grounded outlet. Never ground an appliance with an open coil heating element such as a radiant heater or a toaster. When using a heavy-duty, three-wire appliance whose outside frame case is grounded, never touch any other electrical device.

When changing a fuse, stand on insulating material—a rubber mat or wooden platform—pull the main switch, and use one hand to replace the fuse.

Always use the correct size fuse for the size of wire. Using a 30-ampere fuse in a circuit wired for a 20-ampere fuse may cause the circuit wires to overheat and could cause a fire.

Wiring. Have house wiring checked occasionally for damage to insulation from friction or rodents. Also check cords for worn or frayed spots. Do not run cords under rugs, across registers, near water pipes, or over damp floors. A child could be electrocuted by grasping a defective cord. Children's hands are frequently damp, and moisture is a good conductor.

Never take a radio into the bathroom. Touching it when in water may be fatal. Water and electricity should be kept away from each other. To guard against defective insulation, always purchase appliances that carry the UL seal. Do not immerse electrical equipment in water unless directions for use state that it is allowed. If any appliance has caused even a slight shock, *do not use it again* until it has been repaired. If children are in the family, keep them away from appliances. Select wall outlets that must be rotated through 90 degrees to make contact with current. Then the child cannot stick a metal toy into the outlet and possibly receive a serious burn.

Short Circuits. The cord of an appliance contains two wires, or two groups of tiny wires to give flexibility, the two wires or two groups separated from each other with insulation. The current flows into the appliance through one wire, then through the appliance, causing it to heat, and out the other wire. Frayed cords may cause the insulation on the wires gradually to wear off until the two bare wires come in contact with each other. There is a sudden flash and a fuse is blown because a short circuit is formed. Quite literally a short circuit; instead of flowing through the appliance, the current takes the short path from one wire to the other.

Let's examine the cause of the flash and blown fuse. Remember the equation $E = RI$. Assume the values are

$$120\ (E) = 12 \times 10\ (R \times I).$$

In this equation the *R* of the appliance has been cut out; in contrast, the *R* of the wires is minimal. It is a true equation; therefore *I* approaches *E* in value: a tremendous upsurge of current. Too much current, too much heat. The fuse is melted. Such a short circuit may happen in your electric iron, where the cord is moved back and forth continuously, and where occasionally the cord may be damaged by contact with the hot iron. A short circuit almost invariably results in a blown fuse; the blown fuse prevents a fire.

PROJECTS AND EXPERIMENTS

1. Trace the flow of electric current from the generating plant to its use in cooking bacon in an electric frypan for breakfast. Define terms used.
2. Draw dials to indicate 1,375 kilowatthours.
3. If last month's meter reading was 1,168 kilowatthours, what will be your bill this month, if your rate is $0.05 per kilowatthour?
4. Determine the cost of using the following appliances for the month of April at $0.04 per kilowatthour.
   Frypan, 1,200 watts, 15 minutes per day
   Toaster, 1,000 watts, 20 minutes per day
   Washer, ½ hp, 2 hours a week
   Electric iron, 1,150 watts, 2½ hours per week
   Refrigerator, ⅓ hp, operates ¼ of each 24 hours
   Which appliance costs most to operate? Why?
5. Determine the resistance of the electric frypan in problem 4.
6. Examine nameplates of eight appliances and record data furnished. What safety seals do they carry? Is any special information provided?
7. Visit local generating station. How much is voltage stepped up for different sections of city? What are the local rates?
8. Draw a diagram of a 150 ampere service entrance to house, indicating wire size, main fuse size, kilowatthour meter, circuit break-or-fuse panel, and ground wire.

8. *(Cont.)*

Indicate on diagram the following circuits attached to breaker panel, labeling correct wire size:

3 general purpose circuits
2 appliance circuits
1 range circuit
1 electric dryer circuit (4,600 watts)
1 freezer circuit (½ hp)
1 dishwasher circuit
1 furnace circuit (15 amps)

7 LIGHT

WHAT IS LIGHT. Light from the sun or from a lamp in the home is a form of radiation affecting the retina of the human eye. Only luminous sources emit light; nonluminous bodies reflect the light falling on them. We see largely by reflected light.

Not only is light the result of heat, but it may come from the excitation of the fluorescent powders in fluorescent lamps, from luminous panels used for ceilings and walls, or from night lights in halls and bathrooms which are luminescent.

Light reflected from a body may be regular, spread, or diffuse (Fig. 7.1). Regular reflection is from mirror-like bodies where the angle between the incident ray and the perpendicular to the surface is equal to the angle between the reflected ray and the perpendicular; the incident ray, the reflected ray, and the perpendicular all being in the same plane. Spread reflection also results from light falling on a shiny surface, but the rays are scattered sufficiently that the exact limits of the incident and reflected rays cannot be discerned. In diffuse reflection the scattering is so complete that no individual rays are seen. A ray is the path of light travelling in a straight line.

THEORIES. Two theories for the propagation of light are accepted to explain the various phenomena connected with light: reflection, absorption, transmission, refraction, dispersion, diffraction, interference, polarization. One theory considers light to be transverse, electromagnetic, high-frequency waves, vibrating at right angles to the direction of propagation. The other theory speaks of light emitted and absorbed in tiny bundles of particles called quanta. Quanta are not, however, like particles of matter. Although somewhat different, electromagnetic waves and quanta are interrelated; we speak of quanta of electromagnetic radiation; in fact, a single quantum of electromagnetic energy is known as a photon. A photon moves with the speed of light, but material particles have a speed always less than that of light. The speed of light is approxi-

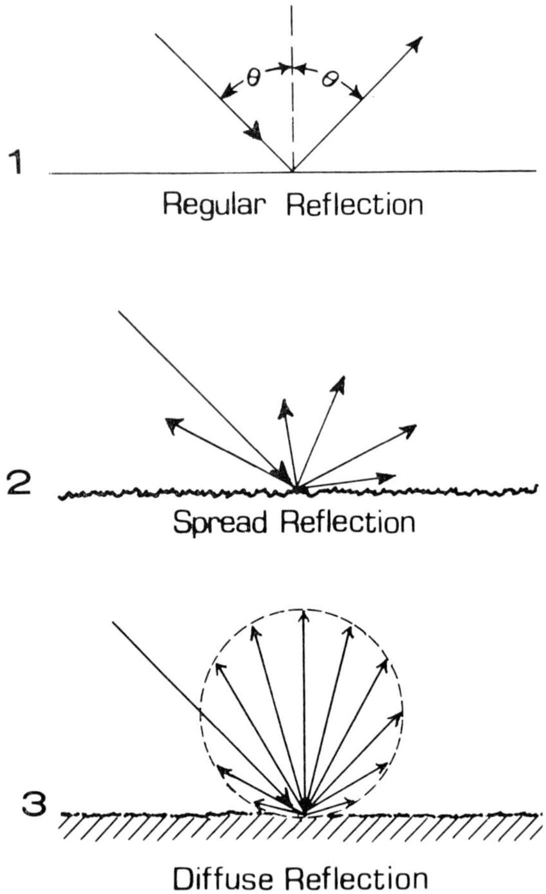

FIG. 7.1. Three types of reflection. (Handbook of Interior Wiring Design, *Industry Committee on Interior Wiring Design.)*

mately 186,300 miles per second in a vacuum, but for all practical purposes may be considered to be 186,000 miles per second. The speed of light through our atmosphere is somewhat less.

PROPERTIES OF LIGHT. When light falls on a surface, some may be transmitted, some absorbed, and some reflected, depending on the nature of the surface. If all the rays are transmitted without a scattering, the material is transparent and an object is clearly visible through the material. If, however, there is scattering of the rays, and the object is seen only as a somewhat fuzzy shadow, the material is translucent. Opaque materials allow no rays to pass. Absorption of light is accompanied usually by the production of heat.

Refraction is the bending of light rays as the light passes from one medium to another, as from air to a liquid, the amount of refraction depending on the type of the two media in contact. Rays are bent toward the perpendicular in entering a denser medium and away from perpendicular when leaving it. Because of refraction, bodies under the water look nearer to the surface than they really are.

WAVE PHENOMENA. Interference, diffraction, and polarization are all wave phenomena. We need the wave theory to account for their action. Interference occurs when light is reflected in spectrum colors from two parallel surfaces of a liquid, such as soap bubbles, a film of oil on water, or occasionally a thin film of muddy water. Diffraction results in a pattern of alternate dark and light lines when light of certain wavelengths is passed through a narrow slit, the pattern depending on the wavelength. Polarization depends on the transverse character of light waves. Edwin Land developed a plastic film, called Polaroid, in which needlelike crystals were aligned in a single direction. Light waves vibrating in that direction are transmitted; light vibrating in other directions is largely absorbed. When two films are used and rotated with respect to each other, light can be dimmed to any desired amount. Polaroid glasses to cut out glare from sunlight are available. More complete discussions of these phenomena are contained in physics textbooks.

To avoid long decimals, scientists have adopted the angstrom unit (symbol Å) to indicate the wavelength of light. One angstrom is equal to 10^8 cm. The visible spectrum extends from 4,000 Å (blue) to 7,000 Å (red). In inches this corresponds to 0.000028 inches (blue) and 0.00016 inches (red).

COLOR. Each wavelength of the visible spectrum produces an individual color, from red of the longest wavelength, through orange, yellow, blue, green, to violet, the shortest. By the use of prisms and gratings (a material containing a large number of parallel slits), or occasionally uniformly spaced lines scratched on glass or metal, light can be separated into these colors. The prism separates white light into its component colors by dispersion. Colors are seen by the light transmitted through them or reflected from them. A transparent sheet of green glass transmits only green wavelength; other colors are absorbed. A yellow dress reflects yellow wavelength, the dress material absorbs the rest of the colors. Other colors than those of the basic spectrum are caused by combinations of wavelengths; brown, for example. When red, green, and blue are mixed, white light is obtained. Colors are also compared as to degree of saturation, i.e., pure colors or mixed with white; red mixed with white gives pink.

MEASUREMENT OF LIGHT. A unit is needed for the measurement of light. Measurements were first made when candles were widely used, so the earliest standard light unit was a sperm candle of a specified size and rate of burning The amount of light given by a horizontal beam from this standard candle was known as candlepower. The light from any candle flame one inch in diameter is equivalent to approximately one candlepower. Candlepower is the measure of light at its source. Later, when electric lamps were used, new light standards were adopted. They were a set of incandescent lamps of known candlepower kept in the National Bureau of Standards in Washington. France and England accepted the same standards so that the fundamental unit was named the "international candle." Now we have a new standard developed in 1948 by the National Bureau of Standards and adopted by the International Committee on Weights and Measures. It is the temperature of solidifying platinum, a temperature of 2,046 K that produces a brightness in the glowing enclosure of the crucible in which the platinum is placed of 60 candlepower per square centimeter.

We are also interested in measuring light on the surface where it is used. It is a basic law that the intensity of illumination on any surface varies inversely as the square of the distance from the light source, when the surface is at right angles to the beam. The law is strictly true only when the source is approximately a point in size. This relationship is expressed in the following equation:

$$\text{Intensity of illumination} = \frac{\text{Candlepower of lamp}}{\text{Square of distance from lamp}} \quad \text{or} \quad \frac{cp}{d^2}$$

In this equation, $cp = \text{foot-candles} \times d^2$, therefore intensity of illuminations is measured in foot-candles.

PHOTOMETER. The amount of light on any surface or the distribution of light around a source may be measured by a photometer. The foot-candle meter is a portable photometer. It has a light-sensitive cell which is independent of the personal error of the user. Light waves from any direction are converted into electric current, the amount of which is indicated on a scale that is calibrated in foot-candles comparable to tungsten filament standard lamps. The sensitivity of the cell is equivalent to that of the eyes so that it gives intensity of illumination in terms of luminous sensation.

A foot-candle meter for home use has a scale from 0 to 75 or 100 foot-candles. Headings above the scale indicate what part of the range is suitable for various tasks. For reading values beyond the scale, a perforated disk may be placed over the cell; the foot-candle values read are then multiplied by ten (Fig. 7.2).

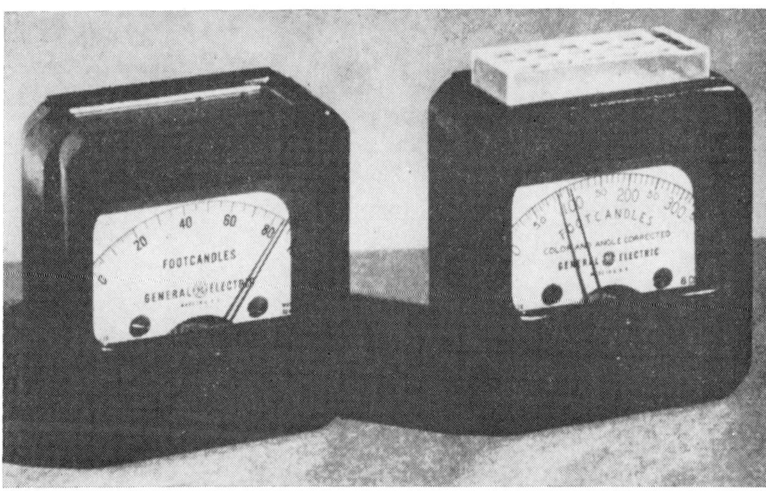

FIG. 7.2. The foot-candle meter measures light at the place where it is used. *(General Electric)*

LUMEN. Light flux, the quantity that can be used, is measured in lumens. If light from a standard unit source falls upon a screen one square foot in area, every point of which is one foot distant from the unit (in other words, part of the inside surface of a sphere), the screen is said to receive a lumen of light. Incandescent lamps commonly used in the home supply from 10 to 20 lumens per watt of energy consumed; fluorescent lamps 60 lumens or more. A photometer is used to measure the lumen output in fixing the rating of any lamp.

LAMBERT. Lambert is the unit of brightness. Brightness of the light source or of a reflecting surface within the line of vision, and the contrast between the brightness of local and general illumination all have an important bearing on visual perception and eye strain. Brightness should, therefore, be measured.

RELATIONSHIPS AMONG LIGHT TERMS. Light terms are usually expressed by letter symbols. It is then possible to write equations to indicate relationships among the different terms. I is the symbol for candlepower; E for illumination in foot-candles; F for lumens; and B for brightness. One such equation is $I = Ed^2$. Since E, foot-candles, is easily read on the meter and the distance d may be measured, I, candlepower, can be determined. Lumens, $F = 4\pi I$. Lumens are measured over a spherical surface, so the factor 4π is used. The area of a sphere is $4\pi r^2$. In this case r^2 is equivalent to d^2 contained in the equation for I. Therefore once you have obtained the value for I, it need only be multiplied by 4π (12.57) to determine the lumens. $B = E \times$ Reflection factor. The reflection factor for any surface is obtained by dividing the foot-candle reading of light reflected from the surface by the foot-candle reading of the light incident on the same surface. To obtain satisfactory results, the surface for which the reflection factor is to be measured should be at least one foot square, and a 16- to 18-inch square is even more desirable.

The foot-candle meter may also be used to obtain the transmission factor of any translucent material by dividing the foot-candles of light transmitted through the material by the foot-candle reading of light incident on the material. The brightness B of the material may be found by multiplying the foot-candle reading of transmitted light, when the cell of the meter is held against the translucent material, by a constant 1.25. This constant is used because the light, incident on the cell almost equally from all directions, is diffused by the material.

The efficiency of a lamp is expressed in lumens per watt. The luminous efficiency of the sun is 80 lumens per watt, of coiled tungsten filament lamps, about 11 lumens per watt, of the fluorescent tube, about 50 lumens per watt.

TABLE 7.1. Comparative Efficiency of Filament and Fluorescent Lamps (Sizes Commonly Used in the Home)

Incandescent lamps		Fluorescent daylight tubes	
Watts	Lumens/Watt	Watts	Lumens/Watt
25	10.4	20	38
40	11.6	30	41
60	13.9	40	45
100	16.2	100	60
200	18.2	...	...
300	19.5	...	...

PROJECTS AND EXPERIMENTS

1. Suggest arrangement of light for a study area; for use of the sewing machine; for playing the piano. Justify choices.

2. Plan adequate lighting for work areas in the kitchen.

3. Compare effect on colors of light from filament and fluorescent lamps. Use duplicate sheets of colored construction paper. Place identical sheets of red, orange, yellow, yellow green, blue, green, blue green, violet, and blue violet under light from filament and fluorescent lamps. Then use samples of wallpaper, carpeting, upholstery material. Evaluate for home use.

4. Compare reflectance of light from various colors. Use sheets of colored paper or pieces of material as in experiment 3. Thumbtack to well-lighted wall. Hold light meter, with sensitive cell toward sample, 2 to 6 inches away from sample, at distance where reading is constant. Record readings and evaluate for application to home situations. A set of colored blotters is recommended; they have mat surfaces that diffuse the light.

8 GAS

GAS used in the homes of the United States may be one of two kinds; natural or liquid petroleum (LP). In other countries where natural gas may not be available, manufactured gas may be used.

MANUFACTURED GAS. Manufactured gas is largely hydrogen and carbon monoxide, as a result of passing steam over carbon (coal or coke).

$$C + H_2O = CO + H_2$$

In practice air must be blown into the retort to maintain the desired temperature and the amounts of carbon monoxide and hydrogen produced are not in a one-to-one equal ratio as Table 8.1 indicates.

When manufactured gas burns, carbon monoxide and hydrogen unite with the oxygen of the atmosphere to form carbon dioxide and water vapor:

$$CO + H_2 + O_2 = CO_2 + H_2O$$

The gas may be enriched by adding other hydrocarbons which supply more carbon and hydrogen atoms.

NATURAL GAS. Natural gas is a combination of methane (CH_4) and ethane (C_2H_6) which, with oxygen, also form carbon dioxide and water vapor:

$$CH_4 + 2O_2 = CO_2 + 2H_2O$$
$$2C_2H_6 + 7O_2 = 4CO_2 + 6H_2O$$

TABLE 8.1. Comparative Heat Value and Composition of Fuel Gases

Gas	Btu/cu ft	CO	H_2	Methane, Ethane	Propane, Butane
				%	
Manufactured	500–550	6	58	27	0
Natural	1,000–1,100	0.1	0	80	2
Liquid Petroleum (LP)	2,500–3,200	0	0	0	100

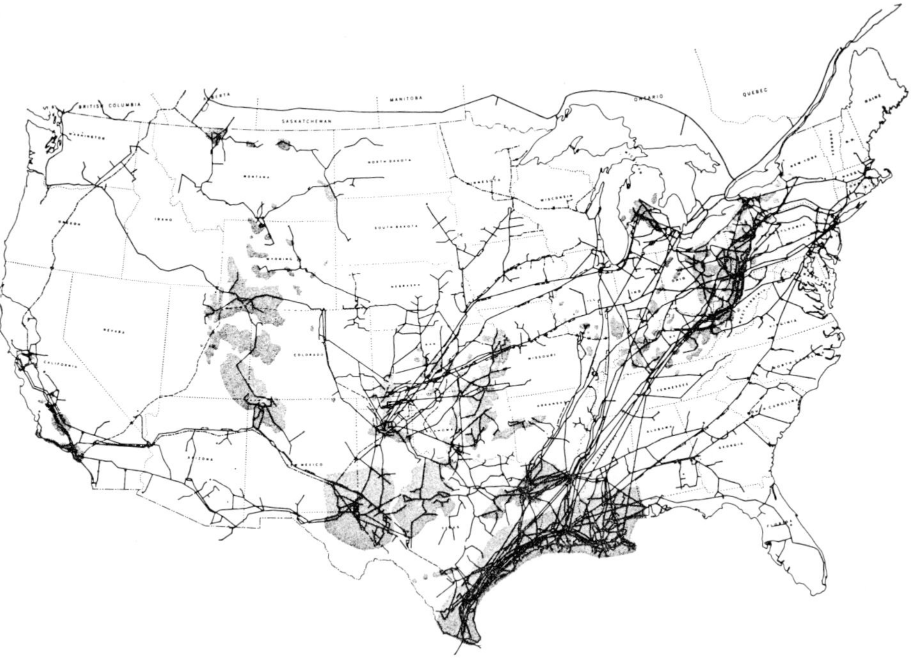

FIG. 8.1. Major gas fields and pipelines. (The Story of Gas, *American Gas Association, Inc., 1970)*

Natural gas is found underground, often as far as five miles below the surface. The gas was formed, perhaps millions of years ago, by chemical action combined with pressure and heat on marine organisms buried in the salt water sands of the ocean floor, an ocean that once covered many continents. The gas was kept from escaping by a thick layer of impermeable rock through which present-day man has drilled. After certain impurities such as gasoline, oil vapors, and moisture, and sometimes sulfur, are removed, the gas is recompressed and transported for long distances through a system of welded pipes that prevent leakage (Fig. 8.1).

LIQUID PETROLEUM GAS. Liquefied petroleum gas is purchased in the form of a liquid under pressure in 47 or 95 cylinders. Rural homes frequently have a 1,000 gallon tank to supply, in addition to home use, the LP gas needed for numerous farm operations. The cylinders are delivered to the home by a distributor; the tank is filled from a tank truck operated by the dealer. The cylinder or tank is connected to the appliances inside the home through a system of pipes and when pressure is removed by the opening of a valve, the liquefied petroleum immediately becomes a gas.

LP gas contains propane (C_3H_8) and butane (C_4H_{10}), again combinations of hydrogen and carbon which burn to form carbon dioxide and water vapor. The propane and butane constituents that make LP gas are obtained from fractional distillation of natural gasoline, from wet natural gas sources, or from oil refineries. They are liquefied and shipped in tank cars to different areas for distribution. Propane is easily changed to a gas under normal temperature conditions but butane may require additional heat to aid in vaporization. LP gas can be used like natural gas in the home.

HEAT VALUES. Gases differ in composition and heat content. The heat value of fuel gases is expressed in Btu per cubic foot. Table 8.1 gives the heat value and percentage composition of the three fuel gases.

The value of Btu per cubic foot varies only slightly from day to day. An Ames, Iowa, report for 30 days, from September 27 to October 26, 1970, indicated the lowest Btu per cu ft value to be 998 on one day and the highest 1,023 Btu per cu ft on another day, but on 23 days out of the 30 the Btu per cu ft value was below 1,010, the average, 1,007 Btu per cu ft. It is interesting to note that the basic price was established at 985 Btu per cu ft, but on every day the heat value was higher.

SPECIFIC GRAVITY. The specific gravity of manufactured and natural gases is lighter than air. At 60 F air weighs 0.0765 pound per cubic foot. Natural gas weighs about 0.054 pounds per cubic foot; manufactured, 0.029 pounds per cubic foot. Liquid petroleum gas is 1½ to 2 times heavier than air, and requires careful handling. If a leak occurs, natural gas will rise and dissipate. In contrast, LP always settles to ground level. If LP escapes within the home from a defective connection, it will settle near the floor. This causes a fire or explosion hazard and is a danger to small children and pets. It will remain below window level, and doors must be opened to remove it from the house.

Manufactured gas, when used, requires some type of storage holder to provide gas instantly on demand. The holders are in two parts, the upper part resting on the gas supply in the lower part, thus maintaining the pressure needed for gas flow. Natural gas is usually stored in the distribution pipes, but sometimes near large cities underground storage holders are provided. Some 200 such man-made caverns for storage of natural gas have been excavated near many of the larger cities of the country, capable of supplying a single day's output of approximately 5 billion cubic feet. Natural gas used in Iowa is piped from Texas fields and is stored entirely in distribution pipes. The pipes vary from ¾-inch to 10 inches in diameter, many of them about 6 inches in diameter. Natural gas is odorless, and an odorant is usually added to help in detecting leaks.

Molecules of gases have much freedom of motion and high velocities, and they tend to fill completely any container, regardless of shape and size. Pressure exerted by the gas in the container is equal in all directions; however, gas will always flow from an area of higher pressure to one of lower pressure. Accordingly, when the valve handle on a gas range is turned to direct the gas into a burner, thus releasing the pressure at that point, the higher pressure in the street main will cause the gas to flow from the main through the meter, and through the connecting pipes in the house system to the range.

GAS METERS. Gas is measured by a meter installed on the outside of the house or where the pipes enter, immediately inside. The home meter consists of two flexible diaphragms which operate on the principle of displacement caused by unequal pressure on opposite sides of first one and then the other diaphragm. When gas flows into one of the chambers, it pushes an equal volume from another chamber. The flow is controlled by sliding valves that transmit their motion to the meter dials. Because of interlocking gears, each alternate meter dial records in the opposite direction, as the dials on the kilowatthour meter do. Record the numbers that the pointers have passed from left to right and add two ciphers to obtain the gas consumption in cubic feet (Fig. 8.2). Sometimes motels have meters into which a quarter must be inserted to cause a certain amount of gas to flow. In boarding houses and foreign pensions in Britain and on the continent most rooms are warmed by gas space heaters operated by prepayment meters.

PRESSURE REGULATOR. In a municipal gas system the pressure regulator, or booster, as it is frequently called, is used to maintain a constant desired pressure in the distributing system regard-

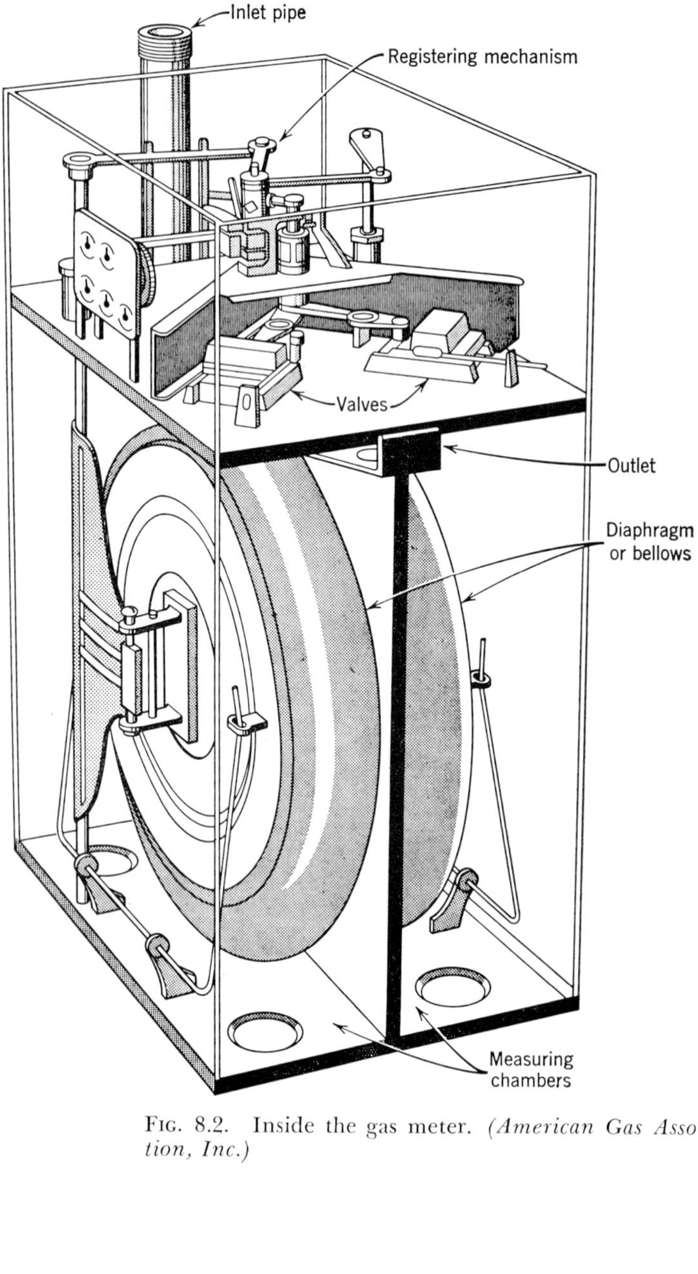

FIG. 8.2. Inside the gas meter. *(American Gas Association, Inc.)*

less of demand load. It can control the pressure in ounces, but in most cities it operates in pounds, often from 1 to 25 pounds. Regulators are installed at various stations throughout the city and they allow the pressure in the distributing system to lower (preventing gas leakage), but respond immediately to load demand.

Occasionally small pressure regulators are installed on individual appliances, the gas furnace and the gas dryer for example. Small units operate on exactly the same principle as the large units, maintaining a constant flow of gas to the burners.

Distributors buy gas from the pipe-line company at a basic temperature of approximately 60 F. If this temperature tends to vary in the local lines, a temperature compensating meter is installed to keep the temperature at the desired level. Large capacity meters, such as are used in manufacturing plants, and hospitals, often must have gas held at a certain temperature. To maintain this temperature, a compensator will be placed in the line.

In the midwestern city of Ames, Iowa (pop. 39,505 [1970]), the gas company buys gas from the pipe-line company at 100 pounds pressure psig (per square inch gauge). A pipe-line system of gas at this pressure forms a loop around the entire perimeter of the city. At the regulating stations at key locations in the city the pressure is reduced to a distribution pressure of 10 pounds psig. At the entrance to a home the pressure is still further reduced to 4 ounces. Gas is burned in the range at 6 inches water pressure, in the furnace and water heater at 3 inches water pressure.

Pressure-recording instruments may also be found at the various stations. They record the pressure data on charts which are checked frequently to note any possible variation in pressure, the cause of which is then investigated. Pressure on either side of the valve is recorded, the gas inlet and gas outlet pressure. Occasionally charts are changed every 24 hours, but most charts are the seven-day type. In Ames the charts are changed each Friday afternoon, in a neighboring town on Mondays, the day depending on the convenience of the company. Figure 8.3 shows that the pressures vary little throughout the course of a seven-day period.

Figure 8.4 represents the principle on which the regulator operates. The actuating diaphragm is mounted on top of the main valve body, its position controlled by the adjustable compression spring above it. If the reduced pressure falls below the predetermined value, pressure under the diaphragm is lost and the spring causes the diaphragm and inner valve to move downward, opening the valve, and allowing high-pressure gas to pass into the reduced pressure area. As the pressure in the reduced area increases, it forces the diaphragm upward, closes the inner valve and decreases the gas flow. This intermittent action tends to maintain a constant gas pressure.

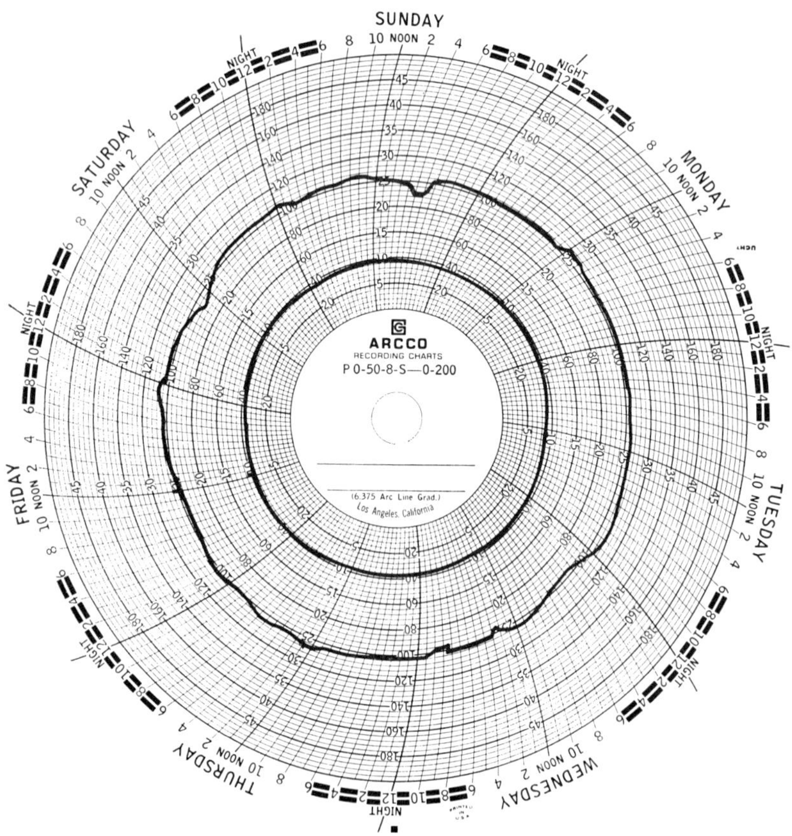

FIG. 8.3. Pressure recording chart from a regulator station. The outer circular curve records the pressure on the high pressure side of the valve, the inner curve, the pressure on the low side. *(Iowa Electric Light and Power Company, Ames, Iowa)*

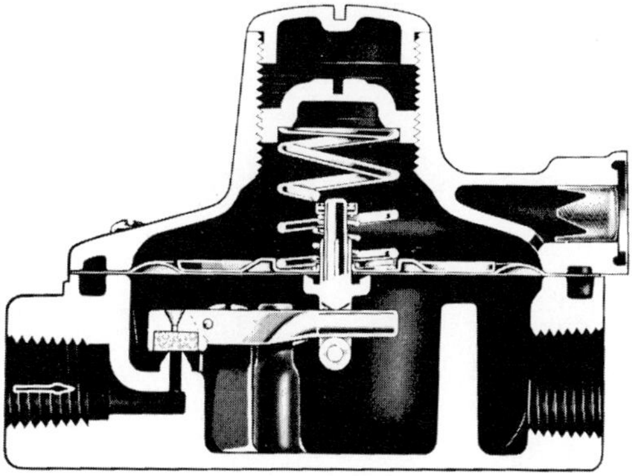

FIG. 8.4. The gas pressure regulator. Fisher Controls. *(Iowa Electric Light and Power Company, Ames, Iowa)*

PROJECTS AND EXPERIMENTS

1. Visit a local gas plant. What is the Btu content per cubic foot of gas? How much does it vary from day to day? From where is the gas obtained? What type of storage facilities are required?

2. At what pressure does your gas company obtain gas from the pipe-line company? Does your city have a perimeter pipe line? What is the pressure in the distribution system at the entrance to homes? Ask to see a pressure recording chart. When are these charts changed? Is a temperature compensator used anywhere in your city? Compare the conditions in your community with those in Ames, Iowa.

3. Read the gas meter for two consecutive months. From the local cost per cubic foot determine what the bill should be.

4. Examine gas appliances in your home. Do any of them have an individual gas regulator? What is its value?

9 HOUSEHOLD ELECTRONICS

ELECTRONICS are no longer limited to the space-age world of science. They have become a part of the everyday life of the homemaker. The phrase, *solid state,* while not always understood, is accepted as a desirable and useful control on a number of household appliances. Solid state controls are similar in function to the vacuum tube, but have the advantages of much smaller size and weight, better automatic control, and greater reliability.

The use of electronic devices began with the development of the vacuum tube. The first radios were made, often homemade, using a galena crystal, a type of lead sulfide. A galena crystal is also a semiconductor, such as solid state components are, but these devices were not understood at that time and so were not used to amplify sound. In fact in order to hear broadcasts, earphones had to be employed. With the development of the vacuum tube sound was amplified and could be heard without earphones.

TRANSISTORS. Then, in 1948, the Bell Telephone Laboratories introduced the transistor, a solid state device that could produce amplifications as tubes do and so usually could be substituted for tubes. Transistors are made from semiconducting materials, usually silicon or germanium. Now, many if not most radios, TV, and stereo sets are transistorized. It may be helpful to define and discuss certain terms before considering these devices in more detail.

Most solid state components are semiconductors, with electrical properties intermediate between conductors, such as copper and aluminum, and insulators, such as rubber and glass.

As noted, tiny crystals of silicon and germanium, often no larger than the head of a pin, form the basis of the transistor. Smaller than tubes, transistors require less voltage, commonly 6 to 12 volts (compared to a hundred or more volts required by tubes), and can be operated from batteries. Both germanium and silicon have the same crystal structure as a diamond. They are stable substances, each atom having four active electrons that are free to bond with four other atoms. They control electric currents without the use of heated filaments or moving parts. Silicon and germanium occur in nature as oxides, which must be reduced and purified before use. Necessary electrical properties are then developed in the crystals by adding to them measured quantities of certain impurities, arsenic to germanium; boron or phosphorus to silicon. The addition of the impurity causes the crystal to have a deficiency or an excess of electrons. An excess of electrons makes the crystal a negative or n-type, a deficiency of electrons produces a positive or p-type crystal. Adding arsenic gives an excess of electrons, an n-type; on the other hand, adding boron causes a deficiency of electrons, and the type is p. The amount of impurity needed is very, very small, about one part in 10 million.

Transistors, diodes, and other semiconducting materials are combinations of n-type and p-type substances. These terms indicate the type of conduction process the semiconductors allow. The n-type means that conduction is by negative electrons; p-type conduction is by positively charged material, the positive charge resulting from the loss of one or more electrons from the atom, sometimes spoken of as a hole from which an electron is missing. The original transistor, a n-p-n combination was used to amplify; the transistor is primarily an amplifying device. P-n-p combinations are also made. A diode is a two-element device, a p-n type. It conducts current more readily in one direction than in the other and is accordingly a rectifier. It is versatile, reliable, of low cost, and can be used to change ac to dc, as in the cordless electric knife. In 1950 a transistor cost $45; it now costs 35 cents.

THERMISTORS. The thermistor is a semiconductor that is a temperature-sensitive resistor. The resistance decreases with increase in temperature. This device prevents an undesirable initial surge of power to a TV set. It also is used in a meat probe to sense the temperature of the roast, and may control the flow of gas or electricity to the oven unit as the roast approaches the selected internal temperature. When the meat has finished cooking, the thermistor may shut off the heat completely, or operate to hold the roast at the serving temperature for a comparatively long period of time.

SCR DEVICE. The SCR device (Silicon Controlled Rectifier) is similar to the diode, but is made of four layers of semiconducting material in the form n-p-n-p. It also allows current flow in one direction only. It must be activated before it will conduct, but once in action, will continue to act until the current is discontinued. This device controls blender speeds, allowing an indefinite number instead of only four or five. Clothes dryers and electric mixers also use this control.

Silicon and germanium crystals are very rugged, but they are sensitive to heat and their use is somewhat limited by the temperature at which they will operate, about 250 F for silicon, but only about 120 F for germanium controls. To prevent the buildup of heat and to dissipate it quickly, a metal jacket, known as a heat sink, is often placed around the solid state item. Moreover, these materials are easily damaged by excessive voltage. For example, if an electric washer, constructed to be used on a 120-volt circuit with a vacuum tube control were accidentally plugged into a 240-volt circuit, the tube might not be damaged. Under the same circumstances, the solid state control would be permanently ruined. In case of failure solid state devices cannot be repaired; they must be replaced. Dust and moisture at junction points may hinder correct operation and these points are usually protected by plastic seals.

PRINTED CIRCUIT. The next step in the miniaturization process was the "printed circuit." The solid state components are assembled on a thin wafer of insulating material and the conducting paths between these elements and any other circuit components, such as resistors and capacitors, are made of thin strips of copper fastened to the nonconducting backing instead of wires. The name printed circuit comes from the fact that the pattern of copper strips is made by a photoprinting process.

The latest addition to these devices is the "integrated circuit." It is a tiny printed circuit made by a microphotoengraving process, in which both connections and components are printed; no soldering of connections is necessary. Several hundred identical circuits may be printed one above the other on a single tiny chip of silicon, layer by layer, until the wanted device is obtained.

It is now possible to go back to a consideration of the use of these controls in radio and TV sets. Radio, stereo, and TV are probably best classified under furnishings rather than equipment, but the homemaker frequently uses a radio in the kitchen while she works, and she does her sewing or ironing with a TV program or stereo music to keep her company.

SOLID STATE APPLICATIONS. To illustrate the value of solid state equipment, the following quotations are taken from RCA advertising leaflets:

RADIOS

Solid state efficiency, with all tubes replaced with transistors, provides undistorted power, instant warm-up and cool operation, no tube burn out.

TV

Long life transistors replace tubes for greater efficiency and reliability. Solid copper circuits replace old-fashioned hand wiring in over 200 possible trouble spots. These circuits of etched copper are permanently bonded to a solid base, will not come loose or short circuit. Solid copper circuits are your assurance of precision manufacture, long life, and dependable performance.

[In the miniature TV that operates from a 12-volt battery] the tubes have been replaced by 21 transistors and a solid state integrated circuit. This tiny chip of silicone, only 1/20 of an inch square, carries an incredibly complex microscopic system of interconnected transistors, diodes, resistors, and capacitors. It provides a new high in TV dependability.

STEREO

A solid integrated circuit puts an additional amplifier right at the sound source. You will not find a single old-fashioned tube in the entire line.

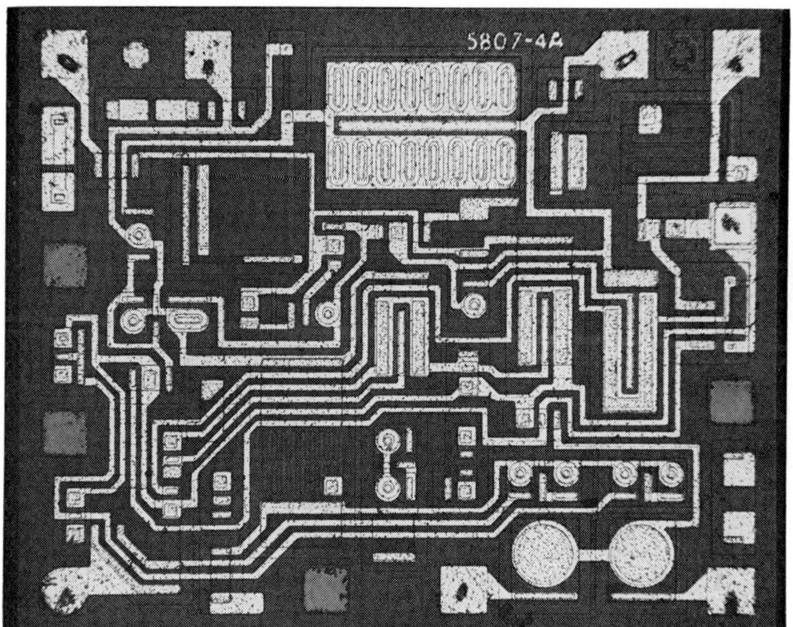

FIG. 9.1. A linear integrated circuit chip. *(Radio Corporation of America)*

They've all been replaced by transistors, a solid state device that runs relatively cool, and less heat generation means greater efficiency—more power output per watt input with low distortion. There is no warm-up wait, the music comes on immediately. Long lived, with no filament to burn out, transistors last indefinitely.

Other manufacturers of radio, TV, and stereo are replacing tubes with solid state components.

When we consider household equipment, we find that solid state devices are widely used. Their reliability, ruggedness, and very small size allow much greater versatility and sophistication than would be possible without them. As semiconductors the crystals have some of the properties of both insulators and conductors, or an in-between property that permits them to allow an electric current to flow, or to prevent its flow as needed.

In their application solid state devices perform a job that could be done, and usually has been done, by electrical switches, but solid state performs at much higher speed. Some applications follow:

1. In clothes washers solid state controlled circuits provide for a selection of a range of speeds for both agitation and spin cycles, and they maintain the speed selected regardless of load conditions, from fast, rugged action for a heavily soiled workload to gentle motion required for hand-wash sheer fabrics, with the possibility of using light-sensitive devices to keep the washing action going until the clothes are completely clean. New washers even provide a soak-wash cycle for woolens, a complete wash and rinse cycle without any agitation.
2. Solid state controls in dryers sense the moisture content of the clothes and stop the cycle when the predetermined degree of dryness is reached.

   A solid state control box in the console panel of the dryer sends a very small current to one of two electrodes on the inside surface of the clothes drum. Since water conducts electricity, the current is conducted from the one electrode through the damp clothes to the other electrode, and back to the solid state control. As the clothes begin to lose their moisture and dry, the amount of current passing through the clothes is reduced. When the preset degree of dampness is reached, the solid state control acts to stop the drying cycle. A few minutes of cool-down time is usually programmed into the control. (Communication from AGA Testing Laboratories, June, 1968)

   The controls are used also to ignite the gas in gas-type dryers, increasing safety and decreasing the cost of a constantly burning pilot.

3. Gas igniters, instead of pilot flames on gas ranges, are a similar solid state control.
4. A defrost control on refrigerators operates only when necessary, thanks to the sensitivity of solid state in detecting frost buildup, thus improving temperature and moisture maintenance, and lowering the cost of operation. Similar components control the self-filling of trays of the automatic ice-cube maker.
5. Vacuum cleaners, floor scrubbers, shampooers, and polishers can be regulated by solid state for any speed from slow to full.
6. The speed of sewing machines is easily regulated by a semiconductor, a four-layer pellet of silicon.
7. Solid state is beginning to be used for the control of furnace heat, by controlling the blower speed with a thermistor so that the blower rotates slowly and gradually fills the ducts with warm air, gently pushing out the cooler air and only reaching maximum operating speed when the temperature has increased sufficiently. In contrast, when a bimetal thermostat is used, the blower when turned on reaches full speed immediately and tends to force the cool air in the ducts out into the rooms at fairly high velocity which may cause a momentary chill, sometimes known as "the cold 70." More definite control of heating and air conditioning systems may be made by determining the temperature and humidity levels at a number of locations in each room by means of solid state, and holding them at a comfortable level by adjusting burners and blowers when necessary.
8. Constant speeds of mixers and blenders can be maintained with solid state elements, regardless of the thickness of the mix.
9. Solid state may perk the coffeemaker.
10. Only with solid state have dimmers for valance, cove, and cornice lighting become practical and easy to operate. A silicon controlled rectifier (SCR), is one type now in use.
11. Finally, solid state has made possible cordless appliances, the shaver, toothbrush, carving knife, flashlight, transistor radio, and recently, the cordless vacuum cleaner, and automatic children's toys—talking and walking dolls and small autos, wagons, and automatic animals.
12. Another well-known application of electronics is the electronic range in which microwaves emitted by the magnetron tube penetrate the food, setting the molecules of the food in motion and so generating heat that brings about the cooking action. It is a rapid process. Microwaves in ranges may have a frequency

of 915 megacycles or of 2,450 megacycles. The microwaves are the same as the hertzian electromagnetic waves used in radio and TV and the United States Bureau of Standards has adopted the term Megahertz (MH_z) in place of megacycle (Fig. 9.2).

PROJECTS AND EXPERIMENTS

1. From specification sheets sent with appliances make a list of those using solid state controls. For what are they used?
2. With the help of your instructor, operate several of these appliances to note how they work. Compare with appliances that do not have solid state controls.

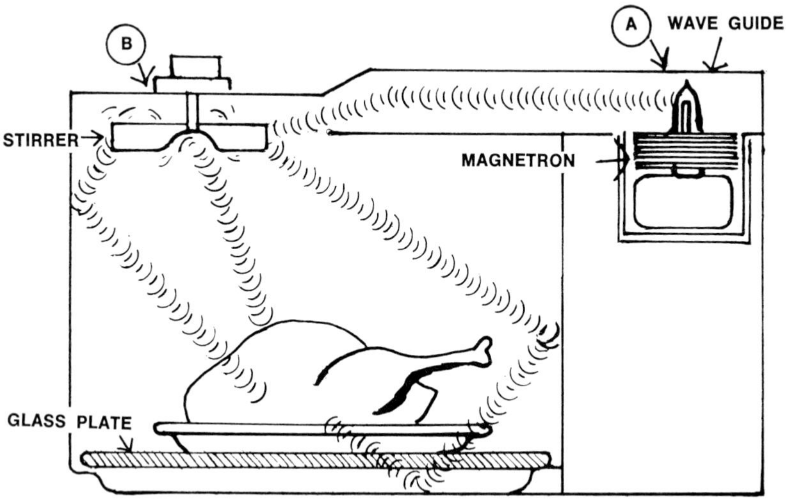

FIG. 9.2. Energy of 2,450 MH_z emitted by the magnetron antenna (A) strikes a slowly revolving fan (B) which sends the power in all directions. It is reflected from walls, back, and bottom of the oven to enter the food from all sides and so cook it evenly. *(Amana Refrigeration, Inc.)*

10 POLLUTANTS

GENERAL ASPECTS OF POLLUTION. Recently much attention has been directed to the problem of pollution in the United States. The focus has been primarily on water and air pollution, although some research has been done on the hazards to health from excessive noise, and the increasing need to do something about the disposal of waste materials, especially since the burning of this material adds to air pollution. Some states have passed laws forbidding this method of disposal.

AIR. Air pollution comes primarily from exhaust fumes from automobiles, buses, trucks, and airplanes. These fumes contain sulfur and nitrogen oxides and carbon monoxide, all injurious to health, and hydrocarbons which cause the smog that has plagued extensive areas of our country. Smog is not only detrimental to health but tends to corrode and erode the city buildings. It adversely affects farm crops and cattle, the resultant property damage amounting to billions every year. Efforts are now being made to produce by 1980 automobiles that will be pollution free, attacking the problem step-by-step in the years before that time. One of the first steps is modifying the engines to use lead-free gasoline. Lead-free gasoline tends to cause deterioration of certain metal parts beneath the hood. The lead in gasoline has acted as a lubricant for these parts, increasing their life span. Lead-free gasoline will cost more; it is more expensive to refine.

Additional air pollution comes from smoke from factories, power plants, and homes throughout the country. Although no direct connection has been proved, pulmonary emphysema deaths in New York City have increased 500 percent in the last ten years. Autopsies show that city dwellers' lungs are blackened, while the lungs of those who live in the nonurban areas of the state are not. Grass and trees are important air conditioners. They temper the sun's heat and provide natural cooling. They maintain atmospheric balance by freeing oxygen. They catch dust, soot, and odors until they are washed into the soil.

WATER. Industrial plants, overloaded municipal sewers, inadequate household septic systems, and farmland and feedlot runoff are primarily responsible for water pollution.

Animal and human waste need oxygen to decompose and use the oxygen in the water for the decomposition process. Partly digested sewage increases the presence of bacteria and the growth of algae in lakes and rivers. At night, when sunlight is not available for photosynthesis, algae use more oxygen from the water for their life processes than the water produces. All too often, after the oxygen demands of decomposition and of algae are met, there is not enough oxygen left for fish and other aquatic life to live, and they die.

> Many standards of water quality set forth in guides or legal codes and regulations are arbitrary and stated to provide minimum rather than optimum support of life in surface waters. Drinking water standards are more stringent and specific inasmuch as water-borne diseases (typhoid, dysentery, etc.) which affect humans are fairly well known. It is anticipated that future research will reveal that water-borne viral-diseases are a greater public health hazard than they are thought to be at present.[1]

Odors and tastes are imparted to drinking water by certain species of algae.

Mercury. Mercury pollution has also become a problem; mercury wastes may not kill the fish, but if absorbed by the fish may render them unsafe for food. Recent reports tell of lakes in certain areas quarantined against fishermen because of this danger. Shellfish tend to store contaminants in much greater concentrations than occur in the water itself. One study showed that oysters stored radioactive substances in concentrations 1,000 times greater than in the water. Contaminated shellfish may cause infectious hepatitis. Other water pollution is caused by drainage from pesticides and chemical fertilizers and from oil spills in coastal waters.

Phosphates. One source of pollution has been phosphates. Phosphates from home and industrial wastes discharged into streams or lakes from sewage plants act as fertilizers, causing algae to grow, a process known as eutrophication. The extensive runoff of commercial fertilizers from cultivated fields during heavy downpours of rain tends to accelerate the process. Detergents also contain phos-

1. Communication from Samuel C. Townsend, University Sanitary Engineer, Iowa State University.

phates, but outstanding independent scientists testified before the Federal Trade Commission that 85 percent of homes in the United States can use phosphate detergents without having any adverse effect upon the nation's waters.

Phosphates can help to nourish the plant life in a lake only when they are present in proper combination with all other nutrients. Phosphates by themselves do not cause eutrophication in any body of water. There are lakes with high phosphate content that show no sign of eutrophication; there are lakes with very low phosphate content with heavy algae growth.

Lake Erie and Lake Ontario are examples of the deleterious effects of pollution. It is estimated that the task of cleaning Lake Erie would cost $40 billion and take a minimum of fifty years to restore the purity it had only twenty-five years ago. But algae were a problem in Lake Erie long before any synthetic detergents were on the market.

Quite naturally, young persons are showing increasing concern for the preservation of the environment. In an exhibition of graphic arts held in New York, more emphasis was given to the destructive effects of pollution than to any other theme. One so-called obituary notice stated that "Lake Michigan, aged 23,031, died recently after many years of abuse stemming primarily from pollution. The lake, once a popular sports and recreation area for millions of people, is survived by Lakes Superior and Huron. Lake Michigan was preceded in death by Lakes Erie and Ontario."[2]

Phosphates have been used in detergents because of their valuable characteristics. They soften water by sequestering the calcium and magnesium hardness ions, they hold dirt in suspension and so prevent redeposition of soil, supply necessary alkalinity and maintain the level of alkalinity during the wash cycle, increase the efficiency of surfactants, and prevent harmful corrosive effects on the washer mechanism.

Detergent manufacturers are attempting to find substitutes for the phosphate ingredients in their products and some companies are advertising new detergents that are phosphate-free. Two such substitutes are silicates and carbonates. They tend to be much less effective in reducing the water hardness and in preventing redeposition of soil. The mineral particles settle back on the fabrics and cause abrasion of the fibers, and they can also cause damage to the mechanism of the washer. Silicates greatly increase the al-

2. *Vista,* March-April, 1970, p. 16.

kalinity of the detergent with adverse effects on some fabrics and on hands coming in contact with the wash water. These detergents should be kept out of reach of young children.

Undoubtedly the best solution to the problem would be to remove the phosphates from the waste water itself. To be wholly successful, the treatment should be carried out in three steps, but often disposal plants use only the first treatment, which allows solid waste to settle out. A second treatment would permit bacteria to act on the solids to decompose them, and a third would use chemical treatment and filtering. It is usually agreed that improved community sewage treatment would solve the matter in the most satisfactory manner. Already technology has been developed for removal of phosphates from sewage and the treatment can be inexpensively implemented.

SOLID WASTES. The disposal of solid waste materials has grown into a national problem. Within the last few years the problem has been increased by the tendency to sell soft drinks, beer, milk, etc., in throw-away containers, and this waste now amounts to about 3½ billion tons a year, or approximately 6 pounds of trash per day for each person in the country. The individual household accumulation is collected two or three times a week by municipal or private carriers and transported to a dump where it is burned in the open or, preferably, covered by landfill. Cities on the coasts sometimes put it on barges that carry it out from the harbor and empty it into the sea, a practice that should be stopped. In a few areas incinerators are used. Incinerators and open burning tend to increase pollution, and in some localities are prohibited.

Public interest in cleaning up the environment has forced numerous manufacturers to share in the effort and to develop new recycling methods of caring for the waste. One company grinds all types of refuse, solid wastes as well as garbage, into fertilizers and soil conditioners. Aluminum and steel companies are collecting discarded cans and melting them down for reuse. Another industry collects discarded bottles and after pulverizing them mixes them with other ingredients to be turned back into new bottles, or manufactured into insulating materials, tiles, or road pavings. Highways paved with these glass compounds are being tried out in some states.

HOUSEHOLD NUISANCES. Our homes are a major part of our environment; therefore environmental pollution problems frequently invade our households.

The air we breathe in the home, on the porch or patio, or in the yard, is affected by the area in which we live, its nearness to manufacturing plants, the closeness of the house to the street, the number of autos and trucks that pass.

Those of us in urban areas expect our municipality to supply us with safe, clear, fresh-tasting water. Rural people are responsible for their own water services and should have them tested regularly. In one Iowa county 200 farm wells out of 1,200 showed pollution by coliform bacteria or by nitrogen or both. Farmers with infected wells must find another source of water, perhaps drill another well. Government insured loans are now available for rural communities to obtain adequate water supplies and to construct waste disposal systems.

NOISE. Acoustics, the science of sound, is of fairly recent origin. It had its beginnings in 1877 when Lord Rayleigh published a treatise, *Theory of Sound.* With the development of radio and other telecommunications under the leadership of the Bell Telephone Laboratories great advances were made, measuring instruments were produced, and terms defined.

It was found that sound, caused by vibrations in an elastic medium, travels in the form of longitudinal waves, in a series of compressions ahead of the wave and relaxations behind the wave, the difference in pressure pulsations creating the sound, which is detected by the human ear. The number of times the pressure changes per second is the frequency of the sound, i.e., the number of complete vibrations per second made by the sound source is the frequency of vibration expressed in cycles per second (cps). The extent of the variation in pressure is measured in microbars. One microbar equals one dyne per square centimeter. A dyne is that force that acting on a body of one gram mass for one second gives it a velocity of one centimeter per second.

The intensity of a sound is its relative loudness. It is measured in decibels. A decibel (db) is a relative quantity, a logarithmic expression indicating the ratio between two sound intensities, one being the reference level, which is usually taken as the pressure of 0.002 dynes per square centimeter, at a frequency of 1,000 cps (taken as the standard tone by the British Standards Institute). For our purpose we shall consider zero decibels as our reference point, using the threshold of hearing as zero. The decibel rating of other sounds will be at varying degrees above the zero.

The lower limit of audibility of a sound is 16 cps, the upper limit 20,000 cps. Audibility varies considerably from person to person and usually decreases with age. Children can hear 20,000 vibrations per second, elderly persons seldom above 15,000. The hearing loss caused by aging is known as presbycusis. Infrasonic sounds are caused by frequencies below those by which the sensation of hearing is stimulated, usually under 16 cps. Ultrasonic sounds result from frequencies above 20,000 cps.

Pitch is the position of a sound on a scale from low to high and depends on the frequency of vibration: low frequency, low pitch; high frequency, high pitch.

Speech is produced by a stream of air forced by the lungs through a slit between the two vocal cords. As the vocal cords vibrate, the slit opens and closes and regulates the airstream. The quality of the sound depends on the resonating chambers of throat, mouth, and nasal cavities. Intelligible speech varies between 600 and 4,800 cps. The average intensity in normal speech at a point 9 centimeters from the mouth is about 7 microwatts per square centimeter.

Hearing is the subjective response to sound by the ear. Sound in the air acts on the eardrum. The motion of this membrane is picked up by three small bones in the middle ear and transmitted to the membrane at the end of the cochlea in the internal ear. The cochlea contains a fluid that passes on the sound to nerve endings, and the nerves carry the impulse to the brain where it is translated into the proper concept.

Sound decreases with distance in open spaces as the square of the distance, but if confined, the falling-off tends to diminish. The velocity of sound increases with temperature and is greater in solids than in air. All sounds have approximately the same speed in air; at room temperatures about 1,130 feet per second. Speed increases 2 feet per second for each degree above 0 C.

Sound is divided into two classifications, musical sounds that have a recognizable pitch, and noise, lacking in pitch, or at least a well-defined pitch. Noise is also defined as any undesirable, unwanted sound, intermittent and erratic and of an infinite number of tones.

Noise annoyance from outside the house comes from passing cars, trucks, buses, industrial machinery, air conditioners, and jet engines of airplanes. It is estimated that the supersonic jets of the future will create a sonic boom of 150 db at ground level. As such a jet flies over the United States from coast to coast this terrific noise from a single plane could affect 20 million persons. Noise above 85 db is regarded a medical hazard that can affect the blood pressure, heart, lungs, and nervous system. Manufacturers of jet engines are working to suppress the noise of these engines.

A truck going by close to a home causes a noise of 90 to 95 db, a loud lawn mower as much as 107 db. The threshold of noise resulting in pain is a loudness of 120 db. One doctor reports that loud noises make tiny blood vessels contract and if frequent enough lengthens the time before they will relax again, leading to injuries to the arteries. Another investigator found that any noise above 70 db caused serious physical disturbances: decrease in flow of gastric juices, dryness of lips and tongue, contraction of chest, abdomen, and leg muscles. Persons who work continuously in noisy areas gradually lose their hearing and may become permanently deaf. Approaching noises are louder than receding noises.

Within the home noise is caused by the equipment used, the vacuum cleaner, washer, mixers, dishwasher, toys of children, radio, TV, record players, air conditioners, and conversation, especially when carried on from one room to another in loud tones. The noise from rock and roll music when turned on high can reach 120 db and can easily affect the sensitive ears and nervous systems of children and young adults.

How can these noises be lessened and controlled? Storm sashes can be left on many windows not needed for air circulation, heavy carpets will deaden noise from the speaking voice and from children playing, insulation in walls and ceilings and in kitchen cabinet doors will prevent the passage of sound. Noise from appliances can be lessened by placing them on rubber or plastic mats, air conditioners can be insulated, squeaky door hinges can be oiled, toilet flush tanks can be adjusted to fill slowly and therefore more quietly. Other plumbing noises can be reduced by wrapping the pipes to prevent their touching the house structure. If possible, use noisy equipment in sound-insulated rooms. Avoid making noise; that is easier than trying to eliminate it. Another method is to cover up, wholly or partially, one sound with another. This is called masking the unwanted sound.

We can try to talk more softly and limit what we say. Quiet periods help to maintain serenity in a home.

ODORS. In earliest times odors were usually spoken of as pleasant or unpleasant, but investigators who were interested in the subject used a series of descriptive terms, 7 to 9 in number. Table 10.1 summarizes the similarities and differences in terms selected. The

TABLE 10.1. Some Descriptive Terms from Linnaeus Onwards

Linnaeus (1752—Botany Materia Medica)	*Zwaardemaker* (1895—Physiology and Perfumery)	*Ammore* (1952/62—Physical Chemistry)	*Schutz** (1964—Psychology & Food Science)
Aromatic	Aromatic	...	...
...	Ethereal	Ethereal	Etherish
Fragrant	Fragrant (floral as sub-group)	...	Fragrant
...	...	Floral	...
...	...	...	Sweet
...	...	Pepperminty	...
...	...	Camphoraceous	...
...	...	...	Spicy
...	...	...	Oily
Ambrosial	Ambrosial	Musky	...
...	Empyreumatic (or burnt)	...	Burnt
Alliaceous	Alliaceous	...	...
Hircine (goaty)	Hircine	...	...
...	...	...	Sulphurous
...	...	...	Rancid
...	...	Pungent	...
Foul	Repulsive	Putrid	...
Nauseating	Foetid or Nauseous	...	...
...	...	...	Metallic

Source: R. Harper, E. C. Bate Smith, and D. G. Land, *Odour Description and Odour Classification* (New York: American Elsevier; London: J. and A. Churchill, 1968), p. 21.
* Schutz uses Pleasantness, Unpleasantness, and Intensity as separate dimensions.

men quoted came from a variety of backgrounds and lived in quite widely separated periods of time and yet seemed to agree on many of the odor classes.

In 1927 Crocker and Henderson selected four basic terms, fragrant, acid, burnt, and caprylic, some part of which they believed all odors possessed, and made a series of samples of each odor, varying in intensity from 1 to 8. Against these samples they checked any odor they were investigating. Depending on the percentage intensity of each of the constituents present, the odor was given a four-figure number. At first well accepted, it was later rejected by certain other investigators on the ground that it did not give consistent results. It is, however, generally agreed that most odors are not due to a single ingredient, but usually include several different qualities. Even at the present time there is no definite system for classifying odors. Man seems less sensitive in his awareness of odor differences than many of the lower animals. A human mother can rarely recognize her child by smell as an animal can.

The source of smell is in the olfactory epithelium located in a narrow cleft in the upper part of the nose. The sensitive area occupies about 4 to 6 square centimeters, and a very small amount of odor is needed to activate the nerve. One nostril is frequently more sensitive than the other.

Odors to be perceived must be slightly volatile and be soluble in the tissues of the olfactory area. Prolonged smelling of one odor, so that all the receptor sites are occupied by the molecules of this odor, leads to olfactory fatigue. Loud noises may inhibit the sense of smell. Odors tend to travel downwind and hunters are aware of this and try to approach their quarry from the opposite direction.

Household Odors. Odors in the home come from cooking, laundry operations, bathrooms, waste disposal, and personal body odors.

Every race finds that other races have an individual smell. We smell like what we eat. Persons on a high meat diet tend to have more odor than those whose diet is largely cereal. For this reason the Chinese and Japanese have little odor and think that Europeans and Americans smell "buttery." White persons often complain of the odor of black people, but it is equally true that the blacks object to the smell of whites. Brunets usually have more odor than blonds, the middle-aged more than the young and old, women more than men. Women have to take special care during the menstrual periods. Certain persons have trouble with foot odors.

Undesirable odors are caused primarily by secretions from sweat and sebaceous glands. The sebaceous glands are found in the hair and secrete oil. The sweat glands are all over the body, supposedly 2 million or more in number, and they help to regulate the heat of the body. Everyone is aware how much more one perspires on a hot day. Persons who do not perspire often seem to feel the heat more oppressively than those who do. Under normal conditions a person loses about 700 cubic centimeters per day in perspiration, more on hot days and in hot occupations, frequently more under conditions of anxiety, excitement, or fright. Clothes tend to absorb odors, and later give them off, causing embarrassment and distress. We tend to prevent or mask such odors with deodorants and perfumes. Some deodorants inhibit the flow of perspiration, others mask the odor. Strong perfumes should not be used, just a faint, almost imperceptible fragrance. Cleanliness is tremendously important and daily baths help to maintain proper hygiene. Soaps cleanse and also mask odors. But poor people often do not have access to bathing and laundering facilities and sometimes not even money for soap.

Cooking odors are, more frequently than not, pleasant: the roasting of meat, the toasting of bread, the brewing of coffee. A person coming into the home from outside will say, "What smells so good?" Odors from the cooking of strong-flavored vegetables, onions, cabbage and the like, may be pleasant to some people, disagreeable to others. A kitchen ventilating fan can be used to exhaust such odors to the outside.

Laundry odors may also be removed by a fan, as can bathroom odors, or they can be masked with suitable sprays, but care should be taken that the covering odor is not itself unpleasant. Antiseptic sprays that have no smell are preferable. The immediate elimination of garbage by the use of a disposer is the most satisfactory method, but if this is not available, the garbage can should be scrubbed and chlorinated at frequent intervals.

The present availability of heavy plastic bags to line the garbage can, which may be tied shut when full, tends to eliminate dirty cans and odors and make the job of collection much more pleasant for the garbage man.

Odors vary from house to house, apparently with the living habits of the occupants. It is especially noticeable when a house has been closed for some time; it seems stuffy, sometimes even oppressive. We open doors and windows to let in fresh air. Air conditioning that recirculates the air in the house may also increase odors in a home. Damp basements in summer cause a musty odor that often penetrates to other areas. A dehumidifier is a good investment. Stale odors remaining from tobacco smoking are often listed first among undesirable odors and are best removed by circulation of fresh air, but deodorants are sometimes used.

Occasionally other household odors come from painting, papering, house cleaning, or from the furnishings themselves—textiles, leather, rubber. Some of these odors are pleasant, at least to some people, others may be considered objectionable. Again, circulation of fresh air is a desirable antidote.

HUMIDITY. Comfort in the home depends in part on the control of noise and unpleasant odors, already discussed, but primarily on the control of temperature and humidity. The relation between these two factors is fundamental. If the temperature is desirable but the amount of water vapor very low, the membranes of the nose become dry and irritated, the lips and skin chapped, and lesions may even develop. Moisture will be absorbed from fabrics and the glue and wood of furniture, causing it to shrink and perhaps split. On the other hand, too much humidity prevents the evaporation of moisture from the body cells and so the cooling of the body, and makes one feel oppressed and lifeless. It will also cause moisture to form on windows and to pass through walls, condensing in the insulation, resulting in damage to the exterior paint. The location in the walls of the vapor barrier is important. To be effective, the temperature of the barrier must be above the dewpoint of the warm side of the wall.

When a certain space contains all the water it can hold at a given temperature it is saturated. Relative humidity (RH) is the ratio of the amount of water vapor in the space to the amount that would saturate the space at that temperature. It is expressed as a percentage.

Relative humidity may be measured by means of a psychrometer. It consists of two identical thermometers mounted side by side on a frame that can be whirled in the air by a handle at the end of the frame. The bulb of the dry-bulb thermometer is the conventional type, the other one, the wet-bulb, has the bulb covered with a muslin sheath connected to a wick which is saturated with pure water before the apparatus is whirled. When the instrument is whirled the moisture evaporates from the wet wick causing the temperature to drop; the amount of evaporation will depend on the degree of vapor saturation of the surrounding air. From the difference in readings of the two thermometers the relative humidity can be calculated.

Maintaining comfortable home conditions is accomplished with winter and summer air-conditioning, usually by heating and humidifying the air in rooms in winter, cooling and dehumidifying it in summer. Persons vary in their reaction to heat, cold, and air humidity, but tests show that a large percentage of persons are comfortable at a temperature of approximately 75 F and 30 percent relative humidity with air circulating between 15 and 25 feet per minute. If the temperature increases to 80 F, the humidity should drop to about 15 percent to maintain comfort conditions. In other words, if temperatures are lower, humidity must be higher, and vice versa, at higher temperatures the humidity must be lower. The humidity level in the home should be coordinated with the temperature out-of-doors. The following RH winter values are recommended:

Inside RH %	Outside Temperature °F
15	—20
20	—10
25	0
30	10
35	20

Such levels of relative humidity would not cause frosting on windows. It is recommended that the humidity level be so controlled that no condensation appears on the interior of double pane windows.

The orientation of the house influences its absorption of heat. If living areas face the south, the rooms may be warmed by the winter sun, but extra care must be taken in summer to combat the natural transmission of heat into the rooms. At the beginning and end of the day, when the sun is near the east and west horizons, the sun's rays will shine through windows on these sides, even though the windows are shaded by awnings.

Structural features also determine the ease of maintaining comfortable conditions in the home: the age of the building, window area, cracks around doors and windows, presence or absence of storm windows or weather stripping, kind and amount of insulation. Heat engineers are trained to calculate the heat loss from a house by finding the average minimum outdoor temperature for the given locality from the reports of the United States Weather Bureau. This is known as design temperature, and is on record for the major cities of the United States and Canada.

As noted, heat comes into the home from outside and also is caused by various housekeeping processes carried on inside. Certain conditions will help to reduce the outside heat. If trees shade the roof and walls of the house, less heat from the sun will penetrate. Using awnings over south windows shuts out sunlight; be sure that the awnings are open on the sides to allow circulation of air. Draw draperies or close Venetian blinds on the sunny side during the middle of the day. Roofs and siding painted white tend to reflect the heat. Louvered openings in the gable ends of attics and an attic fan to push out the heat and draw in cool night air helps in keeping the rooms below the attic cool. Insulation in walls, below attic floors, and above basement crawl spaces lessens the transfer of heat in winter.

Heat inside the house comes from cooking operations, from laundry, and from bathing. Install a kitchen fan and when it is in operation, close the doors to other areas of the house and open a window a short distance to allow ventilation. The fan will also remove excessive humidity. When taking a bath, close the bathroom door and open a window. Do not hang wet clothes in the bathroom or basement to dry. They tend to increase the humidity. Use a ventilating fan in the laundry and vent the automatic dryer to the outside of the house.

Many furnace systems are equipped to add moisture to the heated air in winter, and to dehumidify the air in summer as it is cooled. If such equipment is not available or proves inadequate, humidistat-controlled humidifiers and dehumidifiers are on the market and may be purchased. The size of humidifier needed is based on the volume of space to be treated, and on evaporating one pound of water per hour, when the inside temperature is 72 F and the inside humidity 35 percent, with outside RH 50 percent. Approximately 1,000 Btu of heat are required to evaporate one pound of water.

A dehumidifier is especially recommended for maintaining a dry basement during the muggy summer months. It is also valuable if heavy downpours cause a certain amount of flooding in the basement. Since it is electrically operated, the dehumidifier itself and its connecting cords should be placed above any contact with the wet floor.

The presence of plants in the home is another source of humidity, not large in amount, but sufficient to be valuable. The water vapor, caused by the respiration of the plants, is independent of the level of watering they receive.

SOLID WASTES. Disposal of solid wastes in the home is a continuing problem for householders. The situation is being helped by the recent development of trash compactors.

Companies active in the manufacture of household equipment have been developing such compactors for the home (Fig. 10.1). The compactor will take the average week's amount of refuse for a family of four: paper, cardboard, glass bottles, metal cans, plastic containers, and garbage, that would fill four conventional garbage cans and would weigh 130 to 140 pounds. This weight gives only a partial idea of the volume, for bottles, cans, and cartons are bulky articles. The compactor will take all this material and reduce it to a block 1.5 cubic feet in size, about $\frac{1}{4}$ the original volume.

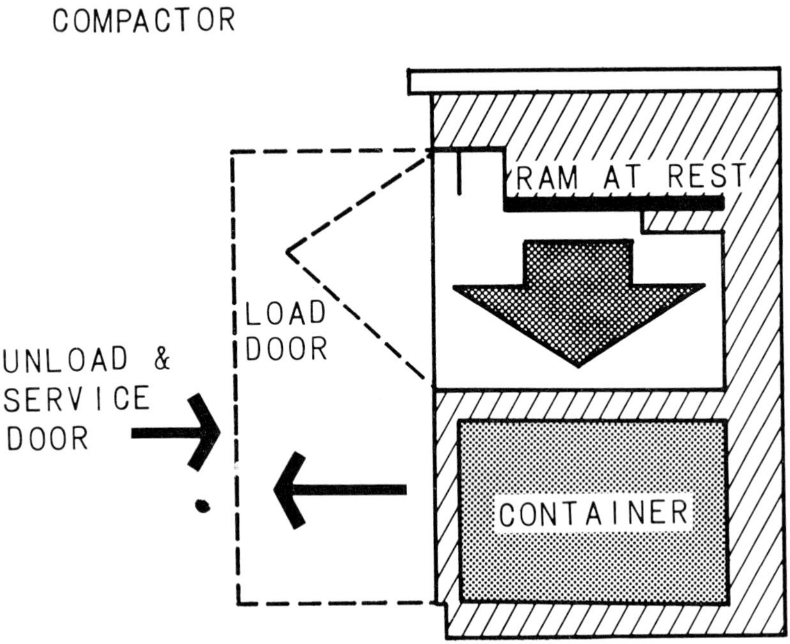

FIG. 10.1. The trash compactor. *(In-Sink-Erator Division, Emerson Electric Company)*

The apparatus is about the size of a single kitchen base cabinet. A rectangular shaped bin for the refuse is mounted on glides so that it can be pulled out and pushed in like a drawer. One container on the market is lined with two layers of wet-strength kraft paper and an inner layer of polyethylene plastic. A 1/3 horsepower induction motor applies a force of approximately 2,500 pounds to a vertically operated ram that descends and crushes and compacts the load.

A safety control prevents operation unless a key turns on the switch. Trash reduced to its minimum possible volume greatly lessens the work of the trash collector and also decreases the community area needed for waste disposal.

In its August 17, 1970 issue, the *United States News & World Report* summarized the cost of eliminating air, water, and land contamination over the next five years: $13 billion for air, $54 billion for waterways, and $4 billion for solid wastes: a total of $71 billion. The map (Fig. 10.2) accompanying the article shows

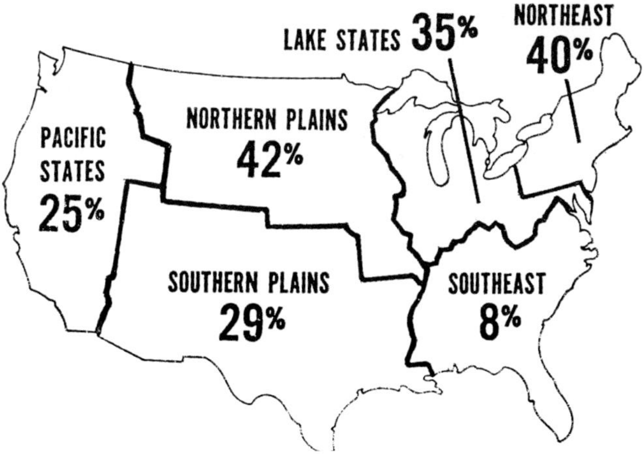

FIG. 10.2. A look at water pollution across the U.S. (U.S. News & World Report, *August 17, 1970, p. 42)*

the percent of miles of polluted streams in various areas of the country. But, as is stated, "Cleaning the environment actually will save the country vast sums, in addition to enhancing the quality of day-to-day living."

It is encouraging to note that university faculties are increasingly interested in the environmental problem and are forming groups with concerns in various areas of the subject—air and water pollution, toxicology, ecology, sanitary engineering, environmental health, urban crowding, misuse of pesticides, noise, etc.—with the objective of improving the environment to preserve the health of man and the quality of living. Such a group can work effectively in the areas of education, research, and service.

Pollution problems are not limited to the United States. London, Paris, Geneva, Tokyo, Mexico City, Rio de Janeiro, Santiago (Chile), and Lima (Peru) all report air and water pollution, and efforts being made to control them.

The effort to minimize pollution is not the responsibility of any one group. We have all shared in the cause; we must all work together to find a solution to the problem.

PROJECTS AND EXPERIMENTS

1. With the use of a Sound Level Meter (made by the General Radio Company, Type 1551-A) compare the noise made by the mixer or by

a sewing machine in a room with the floor (1) of uncovered wood, (2) covered by linoleum or vinyl, (3) covered by a carpet.

2. Notice the difference in intensity of sound of an auto passing the house, when it is approaching and when it is receding.

3. Test your hearing in each ear holding a small wrist watch at 1, 4, 8, and 12 inches from the ear.

4. Classify five cooking odors as pleasant or unpleasant.

5. Rate odors from five bottles of different perfumes on basis of desirability. Compare results of ratings by different members of the class.

6. If possible borrow a sling psychrometer from the physics department and determine the RH in the laboratory by means of the wet- and dry-bulb readings and the use of a psychrometer chart. Do the same for other rooms in the building, an assembly hall, a reception room, a teacher's office.

7. With the use of a portable dehumidifier, operated in a humid room for six hours, collect the moisture absorbed. Determine the volume of air in the room and figure the absolute humidity.

8. Ask a representative from a local distributor to demonstrate the trash compactor.

REFERENCE SOURCES

MAGAZINES

American Dyestuff Reporter
American Fabrics
Appliance Manufacturer
Changing Times
Chemical Engineering
Consumers' Bulletin
Consumer Reports
Detergent Age
Home Furnishings Daily
Home Furnishing Review
Housewares Review
Illuminating Engineering
Journal of Home Economics
Light and Lighting
Merchandising Week
Modern Textiles
Soap and Chemical Specialties

MANUFACTURERS

American Radiator and Standard Sanitary Corp., 40 West 40th St., New York 10018
Armstrong Cork Co., 1010 Concord St., Lancaster, Pa. 17604
Calgon Corp., 1965 Calgon Center, Pittsburgh, Pa. 15219
Colgate-Palmolive Co., 302 Park Ave., New York 10022
Culligan, Inc., South Sherman Road, Northbrook, Ill. 60062
Dow Chemical Company, Midland, Mich. 48640
E. I. DuPont de Nemours & Co., 1007 Market St., Wilmington, Del. 19898
Formica Corporation, 4632 Spring Grove Ave., Cincinnati 45232
Harper-Wyman Co., 930 N. York Rd., Hinsdale, Ill. 60521
S. C. Johnson and Son, Inc., Racine, Wis. 53401
Lever Bros. Co., 390 Park Ave., New York 10022
Owens-Corning Fiberglas Corp., Fiberglas Tower, Toledo, Ohio 43601
Proctor & Gamble Co., Ivorydale Technical Center, Cincinnati 45217

Robertshaw Controls Co., P.O. Box 489, Irwin, Pa. 15642
Warwick Electronics, Inc., 7300 N. Lehigh, Chicago 60648
Wiremold Co., Hartford, Conn. 06110

ASSOCIATIONS

American Association of Textile Chemists and Colorists, P.O. Box 12215, Research Triangle Park, N.C. 27709
American Gas Association, Inc., 605 Third Ave., New York 10016
American Gas Association, Inc., Laboratories, 8501 East Pleasant Valley Road, Cleveland (Independence), Ohio 44131
American Heart Association, 44 East 23rd, New York 10010
American Home Appliance Manufacturers (AHAM), 20 N. Wacker Drive, Chicago 60606
American Home Economics Association (AHEA), 1600 Twentieth St. N. W., Washington, D.C. 20009
Home Economics in Business (HEIB), 1100 Seventeenth St. N. W., Washington, D.C. 20036
American Institute of Architects, 1735 New York Ave. N. W., Washington, D.C. 20006
American Iron and Steel Institute, 150 East 42nd St., New York 10017
American National Standards Institute (ANSI), 10 East 40th, New York 10016
American Society of Heating, Refrigerating, and Air-Conditioning Engineers, United Engineering Center, 345 East 47th St., New York 10017
Bakelite Company, 30 East 42nd St., New York 10017
Better Light Better Sight Bureau, 750 Third Ave., New York 10017
Consumers Union, Mt. Vernon, New York 10550
Detergents, Inc., 1147 Chesapeake Ave., Columbus, Ohio 43212
Edison Electric Institute, 750 Third Ave., New York 10017
Electrical Womens Round Table (EWRT), Ruth Whiting, Executive Secretary-Treasurer, 5630-D Coach and Four Drive, West Dayton, Ohio 45440
Electronic Industries Association, 331 Madison Ave., New York 10017
Equitable Life Assurance Society of the U.S., 393 Seventh Ave., New York 10001
Federal Trade Commission, Washington, D.C. 20580
Gas Appliance Manufacturers Association (GAMA), 60 East 42nd St., New York 10017
Household Finance Corporation, Money Management Institute, Prudential Plaza, Chicago 60601
Illuminating Engineering Society, United Engineering Center, UN Plaza, New York 10017
Industry Committee on Interior Wiring Design, Room 1650, 750 Third Ave., New York 10017

Institute of Electrical and Electronics Engineers, Inc. (IEEE), 345 East 47th, New York 10017
National Better Business Bureau, Inc., 230 Park Ave., New York 10017 (also local offices)
National Bureau of Standards, Washington, D.C. 20234
National Electrical Manufacturers Association (NEMA), 155 East 44th, New York 10017
National Housewares Manufacturers Association, 1130 Merchandise Mart, Chicago 60654
National Wiring Bureau, 155 East 44th St., New York 10017
Plastic Housewares Manufacturers Association, 5719 W. Chicago Ave., Chicago 60651
Porcelain Enamel Institute, Inc., 1145 19th St. N.W., Washington 20036
Small Homes Council, Mumford House, Univ. of Illinois, Urbana, Ill. 61503
Soaps and Detergent Association, 295 Madison Ave., New York 10017
Society of the Plastics Industry, Inc., 250 Park Ave., New York 10017
Superintendent of Documents, U.S. Government Printing Office, Washington, D.C. 20250 (Write to this address for most of the USDA publications. Send cash or money order; do not send stamps.)
Underwriters' Laboratories, Inc., 207 East Ohio St., Chicago 60611
Vacuum Cleaner Manufacturers' Association, 1070 East 152nd St., Cleveland 44120
Water Conditioning Foundation, 120 Waukegan Road, Glenview, Ill. 60025

INDEX

Z